AGRICULTURAL GROWTH AND EQUITY

AGRICULTURAL GROWTH AND EQUITY

(A MICRO-LEVEL EXPERIENCE)

C.H. SHAH
VIMAL SHAH
SUDARSHAN IYENGAR

Foreword by
PRAVIN VISARIA

CONCEPT PUBLISHING COMPANY, NEW DELHI 110059

The Indian Council of Social Science Research has given a financial grant towards the publication of this work. However, the responsibility for the contents of the book — facts, opinions, conclusion etc., is entirely that of the authors and not of the ICSSR.

ISBN 81-7022-358-X

First Published 1991

Published and Printed by
Ashok Kumar Mittal
Concept Publishing Company
A/15-16, Commercial Block, Mohan Garden
NEW DELHI 110059 (India)

Lasertypeset by
Microtech Advance Printing Systems (Pvt.) Ltd.
H-13, Bali Nagar
NEW DELHI 110015

FOREWORD

LONGITUDINAL STUDIES are widely recognised as the best way to investigate the process and problems of socio-economic change. One of their strengths is the obvious possibility of identifying the temporal sequence, which is presumed to improve the chance of discerning the cause-effect relationships that must otherwise be inferred from correlational analysis in cross-sectional studies. Unfortunately, longitudinal research also tends to be slow, costly, taxing for the researchers (who are often pulled to other themes by varied pressures and other developments), and difficult because of the inevitable need to sift evidence on numerous concomitant processes at work.

A short cut to longitudinal research is, therefore, sought through a cross-sectional study of the same population at two points of time, an approximation to the "comparative statics" approach often used in Economics. Such a quasi-longitudinal study has been the great strength of the Second Resurvey of Matar Taluka, which was an ambitious attempt to document the impact of spread of irrigation in an agriculturally developed district of Gujarat State. The resurvey in 1975-76 traced 863 of the 998 households that were interviewed earlier during 1965-66, along with their 188 off-shoots and an additional sample of 60 new households, presumably in-migrants to the 28 selected villages. The imaginative approach of the principal investigators (the two senior authors of this book) led to the collection of a wide range of data and resulted in a wealth of information that posed several intricate problems in analysis and interpretation.

Unfortunately, the usual under-estimation of the costs of elaborate data processing essential for longitudinal research, common all over the world, and inflation slowed down the analysis of the data. The Indian Council of Social Science Research, the sponsor of the Second Resurvey, could not provide more than Rs. 92,000 (considered a generous amount in 1974) for the study.

The pace of work could accelerate only when some economists at the World Bank expressed an interest in encouraging the exploration of the resurvey data to find some pointers to answer the numerous important questions that inevitably arose in their appraisal of the likely impact of the Sardar Sarovar dam to be constructed on the Narmada river. The completion of this book owes a great deal to the stimulus provided by these far-sighted friends in the World Bank.

For the Gujarat Institute of Area Planning, the completion of this book has been a landmark. The second resurvey itself was a cornerstone of the Institute's effort to broaden its research agenda beyond "area planning" to larger issues of development and change. The book bears testimony to the long-standing collaboration between the founder Director of the Institute, Shri Vimal Shah, and Dr. C.H. Shah, initially my teacher and later colleague at the University of Bombay for many years, whose abiding interest in the study has continued even during his sojourns and eventual shift to the U.S.A. Their labours were shared by Dr. Sudarshan Iyengar, who joined the Institute as a Fellow in 1982. Dr. Iyengar shouldered the main responsibility of checking various details for the preparation and revision of two drafts that preceded the final version being published as a book.

The Institute is beholden to all the three authors of this book for their tireless and painstaking pursuit of an unusually difficult method of research with a complex data set. They have done so despite the constant constraint of paucity of resources that continues to plague the social science research institutions in our country.

The Gujarat Institute of Area Planning
Gota 382 481, Ahmedabad, India
July 1990

PRAVIN VISARIA
Director

PREFACE

THE PERIOD from 1965 to 1975 and subsequent years witnessed many major developments. Green Revolution not only changed the perspective for agricultural production, but also influenced thinking of scholars on problems of rural development. The planning process initiated in 1951 in the country reflected these changes, several new schemes were initiated for poverty alleviation and acceleration of growth.

The major problem that persisted without any letting and caused anxiety to policy-makers and researchers pertained to impact of development on equity of income distribution and on poverty. The question repeatedly asked in this context is: Can rural development help reduce substantially the extent of poverty and make distribution of income perceptibly less unequal?

Question persists, but answers to it vary. In the present study we have addressed ourselves directly to the examination of this question. Our study suggests rural development to be a happy correlate of equity and of poverty reduction.

Micro-level experience provides an insight though not final answer to the macro-level situation. Micro experiences do not linearly add up to a macro-level aggregate. Interactions among micro happening could be large, less manageable and adverse. To this knotty problem of interactions we have only incidentally referred. The insight obtained through micro level exercise is still very important. It helps in guiding development forces along the desired lines. The problem of dynamics of the development process which concerns institutions, administration and individuals' behaviour in the context of the market mechanism is complex. To obtain deeper understanding, all related aspects need to be examined. This is too ambitious to be handled through a single project. Dynamics of rural growth are examined by us therefore in a limited context of changes in production, inputs, technology, savings, investment and consumption. Our major contribution is indeed important. We have captured the process of transition of agriculture from subsistence or self-reliance to market related modernisation. Modernised agriculture gives rise to some new markets such as that of services of tractors, machines, migrant labour. The process of modernisation changes also inter-relations of factors of production, inputs and levels of technology, factors substitute each other with increased ease. This process of linked changes has direct and indirect effects on producers, workers and other sections of the society within the taluka, the impact spills over to other sectors of the economy and sections of the community outside the taluka as well. Our findings relate to all these aspects. Our conclusion is that the total effect is not inimical to the social welfare and common good achievable through growth. Consumption level of lower sections of the community improves comparatively faster than that of others.

Authors are grateful to several institutions and individuals for the help received from them at various stages of our study. The comprehensive survey of the rural life undertaken by us would not be successfully carried out and completed but for invaluable and willing cooperation of all concerned.

The Indian Council of Social Science Research (ICSSR) accepted our research proposal for the Second Resurvey of Matar Taluka and provided substantial finances. One of the authors, Shri Vimal Shah was provided a fellowship by the Harvard University Centre for Population Studies to design the study and frame the questionnaire. Later on help was also extended by the Centre to Dr. C.H. Shah, another author, in undertaking analysis in which Dr. Robert Repetto of the Centre was actively involved. The Project was conceived at the time when one of the authors, Dr. C.H. Shah, was a faculty member of the Department of Economics, University of Bombay. The project was conceived therefore as a joint venture and the University readily agreed to permit active participation of Dr. Shah in the project, accepting the recommendation of Dr. D.T. Lakdawala, the then Director, Department of Economics. Late Shri Utsavbhai Parikh, the then Minister of Agriculture, Gujarat Government, took personal interest in the research project. He issued orders to all concerned government offices to extend cooperation to the project. Shri Gordhanbhai S. Patel, Chairman, Gujarat State Land Development Corporation, and a prominent leader of the area, extended all help in the successful organisation of the field work. Shri P.B. Buch and his colleagues, especially Shri Savdasia and Shri Kishore Trivedi in Bureau of Economics and Statistics,

Government of Gujarat, were associated with the project from the start. Evolving a schedule for a comprehensive survey in a manner so that processing involves minimum of labour is a specialised job. Shri Buch and Shri Savdasia not only helped in framing the schedule in a manner that would facilitate putting data on computer, they also helped in putting the data on tape.

Shri P.M. Pandya and Shri Kirti Desai supervised the field-work. Shri Jitendra Joshi, Shri Upendra Upadhyaya and Shri Kamlesh Vyas worked as investigators. They also helped in coding and tabulation. Kum. Meena Gor, Kum. Harsha Shah and Shri Niranjan Bhatt helped in tabulation. Dr. Indira Hirve supervised the transfer of data from schedules to tape. She also supervised the early phase of tabulations.

Shri B.L. Kumar of the Institute helped in tabulation of data. He also supervised the work of statistical clerks. Though Dr. Sudarshan Iyengar joined the project at a late stage his zeal and dedication and research insight helped speed-up the project work. He helped in seeing through the report till the end.

When the Narmada Valley Irrigation Project entered a crucial stage and the World Bank got interested in providing finance in part for the project, the World Bank became directly interested in the Mater Research Project. Aspect pertaining to impact of irrigation on energy, dairy, employment, consumption, and trade links, and return to irrigation and effect of weather on production in irrigated areas interested the World Bank. The Institute undertook the extended research at the instance of and with the finance provided by the World Bank. Resources provided by the World Bank helped us in completing the analysis of the data for the project as well. As a part of the World Bank Project Dr. B.B. Patel undertook the study on labour market, Dr. N.M. Patel undertook the study on energy and dairy, Shri N.C. Shah undertook the study on consumption, Shri B.D. Kumar undertook the study on trade links. Dr. Iyengar examined aspects of costs of irrigation project, and Dr. Anil Deolalikar studied production, income, asset, and investment aspects in terms of household decision-making. These additional research studies enriched the analysis of the main project.

A seminar was organised by the Gujarat Institute of Area Planning to discuss the draft report in February 1987 at Gandhi Labour Institute. It was presided over by Prof. M.L. Dantwala, an eminent economist in the country and was attended by well known economists such as Dr. Y.K. Alagh, Dr. A. Vaidyanathan, Dr. Nilkanth Rath, Dr. Bhupat

Desai, Dr. Ravindra Dholakia, Dr. D.K. Desai, Dr. Mahesh Pathak, Dr. Pravin Visaria, Dr. Tara Shukla, Dr. Viplav Dasgupta, Dr. Ranjit Gupta, Dr. G. Parthsarthy, Dr. Barmera and others. Useful suggestions were made during the seminar. The ICSSR had referred the draft report to two referees according to its usual procedure. They had also made useful suggestions.

All these suggestions were taken into consideration while preparing the final report.

Dr. Pravin Visaria, Director of the Gujarat Institute of Area Planning and Dr. Lakdawala, its Chairman were available for guidance in both research and administrative problems during the project. We received help from Gujarat Vidyapith and Shri Mustali Masavi in particular, at various stages especially in retrieving data for 1965.

The most valuable help we received was from the respondents and the local knowledgeable persons. Not only they willingly provided information which required repeated sessions spread over hours of their valuable time but also they helped us in tracing back a large majority of households studied ten years ago. Shri B.A. Shaikh, Mrs. Sheela Devdas, Mrs. Chandra Krishnamoorthy, Mrs. Girija Balkrishnan and Mr. V. Pandiyan undertook typing for the project. We are immensely grateful to all these institutions and individuals for their help in the project.

Last, but not the least, we are immensely grateful to Dr. Pravin Visaria for his encouraging foreword.

Ahmedabad:
August 1990

C.H. SHAH
VIMAL SHAH
SUDARSHAN IYENGAR

CONTENTS

LIST OF TABLES

LIST OF APPENDICES

1

INTRODUCTION

The Theme

THE STORY of Matar is one of experiences in rural development. It is spread over four and half decades and three field level surveys. Advised by Gandhiji, J.C. Kumarappa, a leading Gandhian economist undertook a survey of rural families of the taluka in 1930. The main objective of the study was to investigate into the heavy burden of land tax on low incomes of peasants. The study also observed underutilisation of human and bullock labour. Between 1930 and 1965-66, when the first resurvey was undertaken, the taluka had achieved progress, mainly through expansion of irrigation. The first survey in 1930 had valuable data on standard of living, on resources used in production and crops. Repeat survey therefore permitted a comparison. The comparison brought out a startling fact, the growth of the economy just matched the pace of increase in population which had doubled. Income distribution however deteriorated, labourers being the main sufferers. Between 1965-66 and 1974-75, the major event of importance for rural development was green revolution. The ICSSR, urged to undertake another resurvey to look into the impact of the Green Revolution.

Though we could contact an overwhelmingly large number, 863 out of 998 households of the 1965-66 sample and added a small fraction, 60 households, to update the sample so as to include newly settled families, the three-point survey may preferably be treated as three snapshots. Where possible we undertake comparison of panel data.

In the present study we confine our analysis almost entirely to two period data, 1965-66 and 1974-75, the former are recast to permit indepth analysis of the process of growth.

Though data-based, the present study is an attempt to follow through a coherent framework of analysis. We prefer to call it the process of marketisation. We shall develop during the course of analysis the concept and later bring together the observations to throw light on it. We give below an outline.

Between 1965-66 and 1974-75, changes occurred faster. The extent of irrigation in 1929-30 was only 6 per cent of the net cropped area. In 1965-66, it was 26 per cent, in 1974-75 it increased to 45 per cent.

The increase in the irrigated area was mainly due to the expansion of canal irrigation. Area irrigated by canals increased from about eight thousand hectares in 1965-66 to about 13 thousand hectares in 1974-75. There are wide annual fluctuations, comparison is therefore only indicative. Use of modern inputs also increased. Typically, fertiliser consumption in the taluka increased from 23,967 quintals in 1965-66 to 31,000 quintals in 1974-75, or by 29 per cent in nine years. The spread of use of high yielding varieties and of tractors, electric motors etc. was more spectacular.

Rapid expansion of irrigation and increased use of modern inputs such as fertilisers, seeds etc. provide a backdrop for the analysis of the forces of growth and their effect on distribution of benefits among different sections of the community. Examining the process of growth and the distribution of benefits are two principal objectives of the study. There were associated changes. Expansion of irrigation in Matar Taluka seems to have initiated what we describe in the study the process of modernisation and marketisation of agriculture. The process of marketisation is described in detail through the analysis of data pertaining to market transactions relating to farm produce, inputs, hiring of durable capital and labour. Production activities usually restricted to the family, become a part of the market when marketisation takes place. Inputs are purchased from the market, their share in the total inputs used, becomes dominant. Labour is hired on wages by all classes of farmers.

Those who do not own improved durable capital, mainly tractors and pumps, hire their services and thus avail of the new facilities. Thus irrigation expansion and high yielding varieties together led to the new phase of production. The new phase covered dairy activities too. Increased market participation on inputs side consequently led to an emphasis on production of crops for markets by all classes of farmers. Irrigation, a major catalytic agent, had thus a pervasive effect on the farm production and on classes, cultivators, labourers and others. In turn, aspects of living which include income distribution, assets holding, consumption, poverty, occupational structure etc. are affected. Our study is a major attempt to observe all these inter-related changes as they emanate from irrigation.

The Sample

The sampling procedure adopted for 1965-66 resurvey consisted of two steps. Firstly, 28 out of the 82 total villages of Matar Taluka were selected for the study. Before selecting villages, the taluka was divided into three natural regions: (i) Navagam, (ii) Matar and (iii) Limbasi. Of them the first two consist of 26 villages each. Nine villages from each of the two regions were selected on a random sample basis. Limbasi has 30 villages. A sample of 10 villages was selected from this region. Thus, in all 28 villages (37.5%) were selected for the study. Secondly, from the sample villages, households were selected for detailed study. House-listing was undertaken on the basis of the occupation of the head of the households. Households were divided mainly into three occupation groups namely, cultivators, farm labourers and non-farm households. Twelve and half per cent of households were selected by random sample from among each size group of holdings. The cultivators were divided into five land size groups: 0-5, 5-10, 10-15, 15-25 and 25 and above acres. From the remaining occupation groups, 25 per cent labourers and non-farm households, were selected.

We had selected a sample of 1059 households in 1965-66. Out of them for 61 household data could not be collected. Remaining 998 households supplied information. Distribution of 998 households according to broad occupational categories is given in Table 1.1.

TABLE 1.1: Distribution of Sample Households (HHs) in 1965-66 and 1974-75

Occupation Groups	Sample HHs in 1965-66	1965-66 sample HHs contacted in 1974-75	Satellite HHs of 1965-66 HHs contacted in 1974-75	Occupation distribution of 1965-66 HHs in 1974-75	Occupation distribution of satellite HHs in 1974-75	New HHs selected in 1974-75	Total sample HHs surveyed in 1974-75
1	2	3	4	5	6	7	8
Cultivators	535	485	134	488	85	5	578
Agricultural Labourers	244	209	31	227	69	18	314
Other Non-farm HHs	219	169	23	148	34	37	219
Total	998	863	188	863	188	60	1111

One of the objectives of the 1974-75 resurvey was to obtain longitudinal data for the households that were surveyed during

1965-66. We, therefore, tried to trace back in 1974-75 as many of households as we could from those surveyed in 1965-66. We found that for 135 households none of the family members could be traced in the village of their original residence; most probably they had migrated lock stock and barrel, a few of them were removed by death or unkind nature. Out of remaining 863 families, 117 branched out. Branched out or 'satellite households' as we term them were 188. We did not take into account of satellites that preferred to settle away from the villages of their residence. During the period of 9 years between 1965-66 and 1974-75 villages under study received new settlers. In order to obtain a representative sample for 1974-75 we selected additional households, adopting the stratified sampling device. Stratification and proportions of the stratified sample were the same as those adopted for 1965-66 survey. Thus, for the 1974-75 resurvey,* the sample consisted of 863 households which had the same heads of households as in 1965-66, or if new head, the same homes. One hundred eighty-eight were satellite households and 60 were new households making a total of 1111 households. Tables 1.1 and 1.2 give details of the sample households and their proportions to the total households in 1965-66 and 1974-75 respectively. Number of sample households surveyed during 1965-66 constituted 15.6 per cent of all households. In 1974-75 their percentage was reduced to 14.

Scheme of Analysis

The core analysis is preceded by an account of major developments based on data obtained at the taluka level during the decade 1965-66 to 1974-75. Chapter 2 gives this account. Chapter 3 begins by tracing changes in the operational land-holdings both among sample households and for the universe from which the sample is drawn. Since 863 sample households out of 1111 studied in 1974-75 were those that were also studied during 1965-66, it was necessary to examine the changes in the occupation among these households. The core analysis begins with Chapter 4, where we examine the development of conjuctive use of wells, tanks and canal irrigation. In Chapter 4 and in Chapter 5, and also in the subsequent chapters, our endeavour is to examine the changes affecting farmers in different size classes of

* For detailed discussion on the nature of 1974-75 resurvey sample refer Appendix 1 of this chapter.

holdings. Chapter 5 studies the process of growth of crop production through the analysis of its three components : (i) area, (ii) yield and (iii) cropping pattern.

TABLE 1.2: Sample Households and Their Percentage to Total Households in 1965-66 and 1974-75

Occupation Groups	1965-66			1974-75		
	Total HHs	Sample HHs	Col. 2 as % of Col. 1	Total HHs	Sample HHs	Col. 5 as % of Col. 4
0	1	2	3	4	5	63.6
Cultivators	4374	535	12.2	4734	578	12.2
Agricultural Labourers	1080	244	22.6	2079	314	15.1
Other non-farm HHs	927	219	23.6	1141	219	19.2
Total	6381	998	15.6	7954	1111	14.0

Use of modern inputs included on the one hand tractorisation, and rapid mechanisation of water lifting. Both these assets had expanded use through increased ownership and increased hiring. Their use directly competed with the use of bullocks. A decline in the use of bullock power helped expand the size of dairy herd. Chapter 6 studies the first part of the triangular process. Chapter 7 examines the spread of use of other modern inputs such as high yielding varieties (HYV) of seeds, chemical fertilisers.

Chapter 8 attempts an analysis of the market behaviour of the producers. Conventionally, the small farmers are believed to be subsistence farmers. They produce crops mainly for family consumption. The present study examines an impact of their exposure to market forces. The findings indicate that the inter-class variations in the extent of marketable surplus seems to have greatly narrowed. This particular process we examine in regard to both the extent of sale of crops and the sale of animal products. The latter aspect is examined in Chapter 9.

Two different reasons, independent of each other, have led to the development of dairy production and animal husbandry as an independent source of growth in Matar Taluka. Increased food production has led to increased supply of dry fodder. Besides, establishment of marketing links for animal products (Matar Taluka forms a part of the Amul network) appears to have helped growth of animal husbandry. Features of animal husbandry activity among sample households in Matar Taluka in 1974-75 are brought out in Chapter 9. This chapter traces also the process of substitution between the dairy and draft animals.

Chapter 10 deals with labour employment and wages. Wage labour is supplied mainly by agricultural labour class and small farmers. Labour contracts are of two types : casual and seasonal or annual. Our hypothesis is that, rapid expansion of irrigation would lead to increased demand for labour, increase being in excess of immediate supply. Rising pressure of demand should lead to an increase in the real wages, However, in Matar Taluka, an inflow of migrant labour followed the expansion of irrigation. Since we do not have a complete account of the supply of and the demand for labour, we examine through the end results like wages and employment, the changes in the equilibrium conditions in the labour market. Our data regarding employment are limited. We get only a feel as to whether there was or there could be widespread unemployment.

Impact of growth is studied through analysis of changes in durable assets and changes in levels of consumption in Chapters 11, 12 and 13. Chapter 11 examines overall levels of assets and debts and investment. Chapter 12 examines levels of consumption. The problem of poverty, its extent during two years of study among different classes, is studied in Chapter 12. Chapter 13 examines consumer durables and also quality of items of current consumption. Together the three chapters bring out the impact of growth through changes in levels and distribution of benefits of growth.

APPENDIX I

Matar Survey Design and its Consequences for Data Interpretation*

In 1965-66 households were selected for study on the basis of two-stage, stratified sample. Two stages were: (1) villages – 28 out of 82 and (2) households. Villages were stratified into three regions on the basis of the physical characteristics. These regions were well defined and are known as (i) Navagam, (ii) Matar and (iii) Limbasi. Selection of village sample was proportional (approximately equal proportion). Households were stratified in three categories i.e. (i) cultivators, (ii) wage labourers working on farm and (iii) non-farm households, on the basis of occupation of the heads of the households. The proportion of the sample was 12.5 per cent of cultivators and 25 per cent from each of the remaining two categories. 1059 households were selected through systematic sampling with a random start. Non-response accounted for 61 of them. Finally 998 sample households were covered.

During 1974-75 the households that were studied in 1965-66 and could be traced back in 1974-75 were included in the sample. The satellites of these households were also included in the sample for 1974-75. Additional sample was selected from new residents in the sample village. They were selected on the basis of the stratification and proportions, same as that adopted in 1965-66.

A total of 7954 households were listed in 28 villages in 1974-75. This list excluded 242 households included in 1965-66 list, but disappeared during the intervening period, either because all the members of the households died or out-migrated. The 1974-75 list contained 354 new households, majority of whom were in-migrants. The remaining households consisted of 1965-66 households that were either undivided or had branched out with satellites in the resident villages. Distribution of total households of 28 villages by status (undivided/divided/new) and by occupation group in 1974-75 is presented in App. Table 1.1 and 1.2. Although we have included 60 new households to the 863 households surveyed during 1965-66 and

* We acknowledge the comments made by Dr. Ravindra D. Dholakia on this subject.

188 satellite households, the group of households studied in 1974-75 cannot be treated as a random sample. Thus, the data that have been analysed is panel data and refers to the behaviour of a particular group that was randomly selected during 1965-66. Insofar as the group is large and covers about 14 per cent of all households listed in Matar Taluka during 1974-75, the behaviour of the group may also indicate the glimpse of changes that have occurred in the taluka between 1965-66 and 1974-75.

APPENDIX TABLE 1.1: Selection of Sample Families in 1974-75

Particular	1974-75				1965-66
	Undivided	Divided	New	Total	Total
Households in 28 sample villages	6272	1328	354	7954	6381
Percentage distribution	78.85	16.70	4.45	100	—
Sample households surveyed	863	188	60	1111	998
Percentage distribution	77.68	16.92	5.40	100	—

APPENDIX TABLE 1.2: Selection of Sample Families by Occupation Group in 1965-66 and 1974-75

Particular	Occupation Category							
	Cultivators		Agricultural Labourers		Households engaged in non-farm occupations		Total	
	1965-66	1974-75	1965-66	1974-75	1965-66	1974-75	1965-66	1974-75
Households in 28 sample villages	4374	4734	1080	2079	927	1141	6381	7954
Percentage distribution	68.61	59.52	16.92	26.14	14.5	14.34	100	100
Sample households surveyed	535	578	244	314	233	219	998	1111
Percentage distribution	53.61	52.03	24.45	28.26	23.34	19.71	100	100

2

A DECADE OF CHANGE

THE ORIGINAL name of the taluka was 'MATBAR' from which the present name 'Matar' is derived. 'Matbar' in Gujarati means prosperous. At one time it was one of the richest talukas of Kheda district. It is an alluvial tract watered by as many as four rivers, namely, the Sabarmati, the Watrak, the Khari and the Shedhi. Since 1963, it also receives waters from the Mahi Canal Irrigation System.

The taluka is situated at the western end of the district surrounded by Ahmedabad district in the west, Cambay and Petlad talukas in the south, Nadiad taluka in the east and Mehmdabad taluka in the north. The taluka has an area of 3,395 sq. km with 82 villages. It is completely a rural taluka having no urban centre. It does not have the benefit of any railway line also.

In the southern part of the taluka *goradu* (sandy loam) soil abounds and in the northern part there is a larger percentage of *Kyari* in (paddy land). Extensive areas are subject to salt efflorescence, particularly near Limbasi. The taluka is divided in three distinct regions namely, (1) Navagam region which has a very large percentage of *Kyari* land and which has been receiving canal waters for nearly a century. It is therefore also known as 'Canal area'; (2) Limbasi region which is a low lying area and was a dry area till 1969; and (3) Matar region having good alluvial soil and having good lift irrigation facilities. Matar and Limbasi regions have been covered by Mahi canal irrigation facilities from 1969 onwards. An attempt is made to take account of the effect of regional features on growth where data permitted.

In this chapter, we deal in detail, with the changes relating to irrigation and the use of modern inputs at the taluka level. We also deal with other changes associated with them. They are related to an increase in population, a change in literacy level, expansion of dairy and cooperative activities. There were a few accompanying problems. A part of the land under cultivation became water logged and a part was affected by salinity. Fairly wide fluctuations, both in the use of irrigation water and in the allocation of area of different crops, were observed. There were a few institutional developments as well. Mainly a descriptive account of these developments with some implications is given here. The account is given at the taluka level, in a few instances developments in three regions are also depicted.

Population

The population in the taluka has continued to grow for the past three decades. Between 1951 and 1961, and between 1961 and 1971 the population of the taluka increased at nearly uniform rate of 30 per cent per decade. The rate of growth of population in the taluka was higher than that in Gujarat State or in Kheda District. During 1961 and 1971, growth of population in the taluka was close to that for the State as a whole (Table 2.1).

TABLE 2.1: **Change in Population — Percentage Growth in the State, the District and the Taluka**

	1951-61	1961-71
1	2	3
Gujarat State	26.9	29.4
Kheda District	22.6	24.0
Matar Taluka	30.5	30.4

The scheduled caste population in Matar taluka constituted 8.5 and 8.7 per cent respectively in 1961 and 1971. The scheduled tribe population increased from 1.5 per cent in 1961 to 2.4 per cent in 1971. This increase was largely due to in-migration of tribal population from neighbouring district of Panchmahals as the employment opportunities in Kheda district in general and in Matar taluka in particular improved due to the expansion in canal irrigation. A natural consequence of growth of population was an increase in the size of villages, villages with population above 1000, increased from 35 to 49.

The proportion of females to males was relatively lower in the taluka compared to that in the State. While in the State for every 1000 males there was 951 females in the rural areas, in Matar Taluka there were 892 females per 1000 males (Table 2.2).

TABLE 2.2: Male-Female Ratio

		Number of females per 1000 males (1971)
1		2
Gujarat State	(Rural Areas)	951
Kheda District	(Rural Areas)	896
Matar Taluka	(Rural Areas)	892

Literacy and Education

In Matar taluka, the extent of literacy was comparable to that in Gujarat State. Between 1961 and 1971, general literacy level increased from 29.5 to 36.6 per cent (Table 2.3) in the taluka.

TABLE 2.3: Level of Literacy for the State, District and the Taluka

	1961			1971			Increase between 1961 and 1971		
	Males	Females	Total	Males	Females	Total	Males	Females	Total
.1	2	3	4	5	6	7	8	9	10
Matar Taluka	41.4	16.2	29.5	49.4	22.3	36.6	7.9	6.1	7.1
Kheda District	48.5	22.7	36.3	55.1	28.5	42.5	6.1	5.7	6.3
Gujarat State	41.2	19.1	30.5	46.1	24.8	35.8	5.0	5.7	5.3

The growth of education is evident from Table 2.4. The table refers to enrolment. Total number of students enrolled in primary school increased from 14,824 to 21,093 or by about 42 per cent between 1965-66 and 1974-75. The number of primary schools increased from 93 to 99. By 1974-75, there was not a single village without a primary school. The total number of teachers — trained and untrained — increased from 388 to 520, or by 34 per cent. Smaller increase in number of teachers would imply a deterioration in teacher/student ratio. However, the number of trained teachers increased at a faster rate, their number went up from 261 to 519 or by 99 per cent. While, nearly one-third of the total teachers in 1965-66 were untrained, in 1974-75 almost all were trained teachers. An increase in the number of

TABLE 2.4: Growth of Primary, Secondary and College Education in Matar Taluka

Year	Primary Education							Secondary Education Students				College Education Students			
		Students			Teachers										
	Schools	Boys	Girls	Total	Trained	Un-trained	Total	Schools	Boys	Girls	Total	Colle-ges	Boys	Girls	Total
1	2	3	4	5	6	7	8	9	10	11	12	13	14	15	16
1965-66	93	9522	5302	14824	261	127	388	9	1385	460	1845	0	N.A.	N.A.	N.A.
1974-75	99	13610	7483	21093	519	1	520	17	2928	1154	4082	0	495	151	646

N.A. = Not Available

Source: Village Schedules, 1974-75.

schools and in the percentage of trained teachers suggest an increased access to the school facilities in the region and a qualitative improvement in teaching. Both these factors may promote rapid growth of literacy in general, and higher level of education in particular in future.

TABLE 2.5: Industrial Classification of the Working Population (Rural), 1961 and 1971

Sl. No.	Industrial classification	Matar Taluka		Kheda District		Gujarat State	
		1961	1971	1961	1971	1961	1971
1	2	3	4	5	6	7	8
1.	Cultivation	53.5	49.5	54.9	56.8	53.3	55.4
2.	Agricultural Labour	26.1	34.2	17.2	27.3	14.8	28.3
	Total (1 + 2)	79.6	83.7	72.1	84.1	68.1	83.7
3.	Mining, Animal Husbandry, Forestry Fisheries etc.	1.2	3.0	0.7	1.6	1.2	2.5
4.	Household Industry	7.7	1.8	6.1	2.1	6.6	2.8
5.	Manufacturing other than household	8.3	1.9	4.8	3.3	6.3	2.7
6.	Construction	2.4	0.9	0.9	0.4	1.1	0.6
7.	Trade and Commerce	2.6	2.9	4.6	2.7	4.8	2.7
8.	Transport	0.7	1.1	1.4	1.2	1.9	1.0
9.	Services	5.2	4.7	9.4	4.6	10.0	4.0
	Total	100.0	100.0	100.0	100.0	100.0	100.0

Source: Census 1961 and Census 1971.

Work Force

We must take note of change which has taken place in the decade — a large decline in the number of persons engaged in household industry. This is an important development. As suggested by the census data percentage of persons engaged in household industry declined from 7.70 in 1961 to 1.78 in 1971 (Table 2.5). If we take together the percentage of persons engaged in household industry and that of persons engaged in industry other than household, we find that the total of the two declined from 8.30 in 1961 to 3.72 in 1971. This decline in non-farm employment seems to have resulted in an increase of agricultural labourers, from 26.10 per cent in 1961 to 34.16 per cent in 1971. The outstanding feature of the changes in the occupational structure between 1961 and 1971 is that the proportion of persons engaged in agriculture (cultivators plus wage labourers) increased sharply. The phenomenon of agriculturisation of rural areas (i.e. increased dependence on agriculture) is a long term feature. The shift in the occupational structure in favour of agriculture during 1961-71 is sharp and unprecedented. The shift in the taluka is a part of the overall process noticed in the district of Kheda and Gujarat State as a whole. The proportion of persons engaged as cultivators and wage earners increased at all levels. The increase was less marked in the taluka. However, the taluka had a much higher proportion of wage earners relative to cultivators in 1961 and 1971. Even if we account for the change in the definition of a worker between 1961 and 1971 censuses, the changes that have taken place in the occupational structure are noteworthy.

Roads and Communication

Between 1965-66 and 1974-75 there was a general improvement in transport facilities in the taluka. Total length of the roads increased from 191 km in 1965-66 to 284 km in 1974-75 (Table 2.6). Metalled roads, however, increased from 153 km to 185 km during the same period. The development regarding roads suggests that while the main lines of communications that were built earlier remained nearly unchanged during the past decade, most of the villages came to be linked with the main system of transport. Increased road facilities help movement of people and the general process of modernisation. More and more people are linked with a wider world. Along with the roads there has been a general improvement in the bus service and the postal and other communication services (Table 2.7). Improvement of transport and communication services has helped the process of integration not only within the taluka but also of the taluka with the outside world. Though there is no institution of higher learning like a college in the taluka, road links provide an easy access to such institutions in adjacent towns. There has been, therefore, a general increase in the levels of higher learning too.

TABLE 2.6: Length of Roads in the Taluka

(in kilometres)

Year	Under Public Works Department			Under District Panchayat			Total		
	Met-alled	Un-Met-alled	Total	Met-alled	Un-Met-alled	Total	Met-alled	Un-Met-alled	Total
1	2	3	4	5	6	7	8	9	10
1965-66	74	—	74	79	38	117	153	38	191
1974-75	109	15	124	76	84	160	185	99	284

Source: Village Schedules, 1974-75.

TABLE 2.7: Communication Facilities in Taluka

(Number of Villages)

Year	Bus		Post Offices		Tele-graphs Office	Telephone Exchange
	All Weather	Fair Weather	Sub-Post Office	Branch Post Office		
1	2	3	4	5	6	7
1965-66	N.A.	N.A.	3	28	1	—
1974-75	43	35	8	33	2	1

Source: Village Schedules, 1974-75.

Social Overheads

Social overheads may pertain to a school, electricity connection, a bus stop, a post office or even to a hospital or a dispensary to which a village has an access. Those interested in agriculture may benefit by the proximity to the village level worker (VLW) headquarters, veterinary dispensary or hospital, cooperative credit society headquarters, an agricultural demonstration farm or centre for sales and purchases and even a headquarter of a village talati. Information on all these facilities was obtained from the village schedules. For convenience of analysis, we have aggregated distance of the village from the location of different facilities. Different facilities are divided into three broad groups, pertaining to (i) communication,* (ii) health,** (iii) agriculture, and other extension services.† On an average for the taluka as a whole, the distance of the communication facilities worked out to 2.73 km. The inter-regional differences were not marked. By and large the taluka had good communication facilities.

Health care facilities were located at an average distance of 9.4 kilometres. Agricultural and other extension services located at an average distance of 7.1 km. There was also a marked improvement in electrification, the number of villages electrified for at least one purpose increased from 15 in 1964-65 to 48 in 1974-75.

Irrigation

In the year 1965-66 irrigation facilities were unevenly spread among three regions. A look at the regional data brings out this unevenness. In 1965-66 the extent of net irrigated area as related to net sown area differed for different regions by a wide margin (Table 2.8). In the Navagam region 43.8 per cent of net sown area was under irrigation, in the other two regions, Matar and Limbasi, the extent of irrigation was 25.9 per cent and 24.8 per cent respectively. In 1974-75 the extent of irrigation increased in the Navagam region to 55.7 per cent or by 12 percentage points. In Matar and Limbasi it increased to 51.9 and 50.6 per cent or by 26 percentage points respectively. Consequently in 1974-75 the inter-regional difference in the extent of irrigation almost disappeared. The rapid expansion of irrigation in Matar Taluka as a whole, was therefore the result of the relatively much greater expansion of area under irrigation in the two regions namely, Matar and Limbasi. In 1974-75, the changing features of the rural economy of the taluka would reflect in one sense, the effect of the rapid expansion of area under irrigation in Matar and Limbasi regions. Hence it may be useful for us to pursue the analysis at regional level to gain an insight into the effect of rapid expansion of irrigation in two regions. In contrast, the rural economy of Navagam would indicate in 1974-75, an effect of full grown, large scale and long-term irrigation. The region

* Communication included distance from high school, Bus Service and Post and Telegraph office.

** Health care facility included the distance from hospital and/or dispensary.

† Agricultural and other extension services included VLW HC, Talati headquarter, Credit Society, Veterinary Service etc.

had reached the maturity and had overcome the difficulties of social and economic gestation involved in utilisation of irrigation facilities. The contrast in the impact of irrigation may be striking and interesting in terms of quality and dependability.

TABLE 2.8: Extent of Irrigation

Region	Net Irrigated Area as a percentage of Net sown Area	
	1965-66	1974-75
1	2	3
Navagam	43.8	55.7
Matar	25.9	51.9
Limbasi	24.8	50.6
Total	30.5	52.1

Note: The data on irrigated area differ for different sources. Aggregation of data drawn from the records of individual villages give the above stated results. The District Level Records suggest that the percentage of area under irrigation to net cropped area was 26 in 1965-66 and 45 in 1974-75. The district records however do not give regionwise data.

Source: Village Schedules 1974-75.

It is important to note that the area irrigated by canals, tanks and wells fluctuated widely till 1967-68 (Table 2.9). For instance, in 1966-67 area irrigated by canals was 4,312 hectares, the lowest for the period between 1963 and 1968. The highest figure of area irrigated by canals was 9,427 hectares in 1965-66. Before 1967-68, the canal irrigation was concentrated largely in the Navagam region. Even in Navagam, only a few villages within a radius of about 16 kilometres had the riparian rights which gave them precedence over other villages in the use of irrigation water. Other villages did not enjoy such rights. After 1966-67, Matar and Limbasi regions benefited largely from the extension of Mahi Right Bank Canal (MRBC) Project. With MRBC Project in operation not only the larger areas in Matar and Limbasi regions benefited, but the Navagam region also experienced a relative stability and dependability of the supply of irrigation water. Since 1968-69, area irrigated by canals fluctuated within a narrow range between 12 thousand hectares (1973-74) and 14 thousand hectares (1975-76).

TABLE 2.9: Area Under Irrigation in Matar Taluka by Sources

Year	Government Canals	Tanks	Wells	Other sources	Net Area irrigated	Percentage of net area sown	Area irrigated more than once	Gross area irrigated	Percentage of Government Canals	Percentage of wells
1	2	3	4	5	6	7	8	9	10	11
1961-62	4722	242	1145	-	6108	14	-	6108	77.3	18.8
1962-63	5654	160	523	491	6828	15	-	.6828	82.8	7.7
1963-64	6997	224	484	308	8013	18	-	8013	87.3	6.0
1964-65	8158	26	861	1112	10157	22	-	10157	80.3	8.5
1965-66	8427	226	1653	1572	11878	26	-	11878	71.0	13.9
1966.67	4312	951	7260	910	13433	35	-	13433	32.1	54.6
1967-68	5062	1363	6075	993	13493	31	1378	14871	37.5	45.0
1868-69	12859	486	2473	386	16204	36	227	16431	79.4	15.3
1969-70	13460	575	2729	338	17102	35	261	17363	78.7	16.0
1970-71	15017	528	2671	1177	19393	46	3144	22537	77.4	13.8
1971-72	13468	491	4577	2575	21111	46	3351	24462	63.8	21.7
1972-73	13998	415	2683	43	17139	38	2833	19972	81.67	15.7
1973-74	11743	865	4986	593	18187	43	3531	21718	64.6	27.4
1974-75	13505	858	4693	160	19216	45	3712	22928	70.3	24.4

Source: Office of the Taluka Development Officer, Matar.

Before 1967-68, area irrigated by wells fluctuated but well irrigation constituted an important source. Since 1968-69, we find a sudden drop in the use of well irrigation till 1972-73. Area irrigated by wells declined to about 2,500 hectares. In 1971-72 however area irrigated by wells increased to 4,600 hectares. Subsequently it has tended to expand once again reaching the level close to 5 thousand hectares.

Details about area irrigated more than once were not available till 1966-67. Double sown irrigated area was as low as 227 hectares in 1968-69. Since then it has risen, it fluctuated between 3 thousand and 4 thousand hectares. Assured supply of water in canals, supplemented by increased use of wells and tube-wells has facilitated more intensive use of land by extending area irrigated more than once. The irrigation developments between 1965-66 and 1974-75 have two major features (i) rapid expansion of area under irrigation which was experienced by Matar and Limbasi regions and (ii) assured supply of irrigation water to all the regions.

Waterlogging

Two problems that emerged consequent to rapid expansion of canal irrigation are waterlogging and soil salinity. In Gujarat, rivers run from East to West, canals run mainly North-South. The rainwater received by the regions flow into the rivers from East to West. The roads and canals together act as blocks to the natural flow of rainwater. In the areas affected by depressions water collects. Besides, liberal use of irrigation water above the prescribed limits also leads to accumulation of water in the low lying areas. It is feared that close to 2,000 hectares of land are affected by waterlogging.* Area affected by waterlogging is deprived of growing *kharif* crop. Land lies fallow. Frequent and more liberal use of irrigation water without drainage facility leads to soil salinity. Additional 2,000 hectares of land is estimated to have been affected by soil salinity.*

As a result of waterlogging and salinity net sown area declined between 1965-66 and 1974-75 from about 46 thousand hectares to about 43 thousand hectares (Table 2.10). There was a consequential increase in land under current fallow of about 1,900 hectares. It may be

* This was communicated to us during our discussions with the officials of the Taluka Development Office, Matar.

inferred on the basis of the land use data that about 2,500 or more hectares of land might have been either adversely affected (waterlogging and salinity) by extension of irrigation or land taken up by roads, canals and other constructions. The damage done to the land is not irreparable but the cost of reclamation would be high. Waterlogged area can be reclaimed only by an extensive drainage system that would cover the entire region including drains in the field.

TABLE 2.10: Land Use in Matar Taluka

Year	Total geographical area	Barren and uncultivable land	Land put to non-agricultural use	Cultivable waste	Permanent pasture and grazing land	Current fallow	Other fallow	Net Area Sown
1	2	3	4	5	6	7	8	9
1965-66	57696	—	8210	370	2766	416	395	45539
1974-75	57407	688	6656	480	3423	2343	560	43257

Source: Annual Season and Crop Reports.

The low lying areas and the land affected by soil salinity are not evenly distributed over different regions. Land affected by waterlogging and salinity is concentrated mainly in Limbasi region. The Government has recognised the seriousness of this problem and it is considering a proposal to establish command area development authorities to tackle the waterlogging and other problems in the command areas of major irrigation systems of the State.

Water Lifting

Between 1965-66 and 1974-75 the number of oil engines used for irrigation increased from 262 to 614 and the number of electric motors from 77 to 454 (Table 2.11). Both oil engines and electric motors are concentrated to a greater extent in Matar and Navagam regions (Table 2.12)

TABLE 2.11: Oil Engines and Electric Motors in the Taluka

(Number)

Year	Oil Engines	Electric Motors
1	2	3
1965-66	262	77
1974-75	614	454

Source: Village Schedules, 1974-75.

TABLE 2.12: Regionwise Oil Engines and Electric Motors, 1974-75

(Number)

Region	Oil Engines	Electric Motors
1	2	3
Navagam	214	157
Matar	223	254
Limbasi	177	43
Total	614	454

Source: Office of the Taluka Development Officer, Matar.

TABLE 2.13: Regionwise and Sourcewise Net Area Irrigated in Matar Taluka 1974-75

(Hectares)

Region	Source of Irrigation				
	Canal	Tanks	Dugwells	Tube wells	Total
1	2	3	4	5	6
Navagam	3554	40	2268	500	6362
	(55.9)	(0.63)	(35.7)	(7.9)	(100.00)
Matar	3463	233	3124	194	7014
	(49.4)	(3.3)	(44.5)	(2.28)	(100.00)
Limbasi	6840	330	979	—	8149
	(83.9)	(4.1)	(12.0)	—	(100.00)
Total	13857	603	6372	694	21526
	(64.4)	(2.8)	(29.6)	(3.2)	(100.00)

Notes: 1. Figures in parentheses denote percentage.

2. The absolute figures quoted above do not tally with the figures quoted in Table 2.15 because the sources are different. However the difference is a narrow one. Figures computed from village records total up to 21,526 acres (gross irrigated area) where as the taluka source gives a figure of 22,928 acres which is 6.5 which is 6.5 per cent higher.

Source: Village Schedules, 1974-75.

Area irrigated by wells was almost as much as area irrigated by canals in Matar region, 3,124 hectares and 3,463 hectares respectively (Table 2.13). In Navagam too, area irrigated by wells was fairly significant, 2,268 hectares compared to 3,554 hectares irrigated by canals. In Limbasi region, well irrigation accounted for 979 hectares but canal irrigation covered 6,840 hectares. It may be inferred that wells were used conjunctively in Navagam and Matar regions, but Limbasi region depended mainly on canals for irrigation.

Access to Irrigation

Agricultural census for 1971 gives detailed data regarding land holdings according to their size and according to whether they are wholly irrigated, wholly unirrigated or partly irrigated (Table 2.14). The data suggest that out of 15,375 holdings, nearly one third i.e., 4,990 holdings were wholly irrigated, but an equally large number of holdings, 4,545, were wholly unirrigated. The rest were partly irrigated. Despite widespread irrigation, nearly one-third of holdings did not have an access to any source of irrigation. This fact would have an implication regarding the pattern of income distribution emerging after the rapid expansion of irrigation. Both wholly irrigated and wholly unirrigated holdings were largely small in size. Partly irrigated holdings were mostly medium and large size holdings. The percentage of area irrigated to the total land held was however, nearly the same up to 10 hectares size of holdings. It declined from 54 per cent for a very small holding of 0.5 hectares, to 45.3 per cent for holdings of 5 to 10 hectares in size. Above 10 hectares, the extent of irrigation declined rapidly for holdings of size 30 to 40 hectares, only 11 per cent of the land was irrigated. Again very large holdings above 50 hectares had relatively larger percentage of area under irrigation. However such holdings were only six.

Crops

Irrigation directly influences cropping pattern. Extension of irrigation has helped expansion of area under wheat and paddy. There is a sharp decline in the area under jowar, pulses and commercial crops like cotton and oil seeds. Paddy needs large and regular supply of water. Canal irrigation does not ensure this to all irrigated holdings. We, therefore, find wide fluctuations in area under paddy from

TABLE 2.14: Distribution of Land Holding by Size Groups and Extent of Irrigation

(Area in hectares)

Size Distri-bution	Total Holdings				Wholly Irrigated Holdings			Wholly Unirrigated Holdings		Partly Irrigated Holdings		
	No.	Total area	Irri-gated area	Col. 4 as % of Col. 3	No.	Total area	Irri-gated area	No.	Total area	No.	Total area	Irrigated area
1	2	3	4	5	6	7	8	9	10	11	12	13
Below 0.5	3111	892	486	54.5	1669	447	440	1246	351	196	94	46
0.5-1.0	2925	2172	1051	48.4	1173	848	834	1099	811	653	513	217
1.0-2.0	3705	5376	2431	45.2	1065	1514	1475	1124	1602	1516	2260	956
2.0-3.0	1974	4854	2212	45.6	474	1152	1110	457	1120	1043	2582	1102
3.0-4.0	1213	4208	1892	45.0	233	799	777	243	846	737	2563	1115
4.0-5.0	770	3440	1571	45.7	149	666	632	149	664	472	2110	939
5.0-10.0	1386	9350	4238	45.3	197	1288	1215	182	1196	1007	6866	3023
10.0-20.0	259	3340	1269	38.0	27	324	309	40	494	192	2522	960
20.0-30.0	22	510	196	38.4	2	47	46	1	20	19	443	150
30.0-40.0	4	138	15	10.9	—	—	—	2	72	2	66	15
40.0-50.0	—	—	—	—	—	—	—	—	—	—	—	—
50.0 above	6	377	131	34.7	1	54	44	2	168	3	155	87
Total	15375	34657	15492	44.7	4990	7139	6882	4545	7344	5840	20174	8610

Source: Agriculture Census, 1971.

TABLE 2.15: Area Under Crops

(Area in Hectares)

Crops	1963-64	1964-65	1965-66	1966-67	1967-68	1968-69	1969-70	1970-71	1971-72	1972-73	1973-74	1974-75
1	2	3	4	5	6	7	8	9	10	11	12	13
Paddy	11533	12704	13041	13463	12857	16227	17760	18684	20104	16626	13363	13235
Jowar (*Kharif*)	1229	1510	1726	1254	2062	2204	2191	1734	1217	1738	1016	546
Jowar (*Rabi*)	52	14	18	35	23	-	30	151	310	194	1086	377
Bajra	8061	6853	8153	8206	8004	4737	9867	8180	8108	9309	10438	11054
Maize	2	4	4	7	17	-	-	33	6	-	-	-
Ragi	812	952	868	1146	1404	1126	389	163	189	148	188	288
Wheat	7500	7146	6915	7066	8693	6475	5967	8981	7590	5981	9259	9229
Barley	-	-	-	-	-	-	-	-	-	-	-	95
Other cereals	571	2583	1517	1199	847	1425	1009	594	3997	1530	695	992
Total cereals	*29760*	*31766*	*32242*	*32376*	*33907*	*32194*	*37213*	*38520*	*41521*	*35526*	*36045*	*35816*
Gram	751	706	212	508	584	630	391	355	541	576	522	279
Tur	1588	1573	1650	1649	1774	1455	1617	894	1652	50	635	302
Other-Pulses	754	120	710	855	929	1121	837	651	783	466	137	149
Total Pulses	*3093*	*2399*	*2572*	*3012*	*3187*	*3206*	*2845*	*1900*	*2976*	*1543*	*1294*	*730*
Total Cereals & Pulses	*32853*	*34165*	*34814*	*35388*	*37094*	*35400*	*40058*	*40420*	*44457*	*37069*	*37339*	*36546*
Vegetables	144	85	58	73	90	117	162	180	327	133	374	394
Fruits	527	713	786	629	271	377	263	384	174	284	303	372

Contd.

TABLE 2.15: *Contd.*

1	2	3	4	5	6	7	8	9	10	11	12	13
Other Food Crops	171	-	102	188	513	102	106	13	159	84	55	71
Total	33695	34963	35760	36278	37968	35996	40589	40997	45157	37570	38071	37383
Cotton	8390	7934	5510	5757	5121	4607	3834	4267	5914	6985	4531	6900
Other fibres	-	16	52	3	6	11	6	3	-	-	-	-
Total fibres	8390	7950	5562	5760	5127	4618	3840	4270	5914	6985	4531	6900
Groundnut	27	25	16	15	-	13	1	-	-	-	-	-
Castor	109	154	188	78	29	134	102	70	81	187	179	174
Sesamum (Till)	-	-	-	2054	742	1189	1363	1571	1000	1336	1248	1310
Rape and Mustard	-	-	-	-	-	-		7	9	87	449	331
Other Oilseeds	2170	2001	1494	3	-	-	2864	14	59	2	8	5
Total Oilseeds	*2306*	*2180*	*1700*	*2150*	*771*	*1336*	*4330*	*1662*	*1149*	*1612*	*1884*	*1820*
Sugarcane	-	8	8	7	3	-	2	3	13	10	3	13
Tobacco	506	1091	866	875	1584	1889	566	373	618	776	448	412
Other non-food crops	4093	3097	5315	5391	755	2502	5	2723	2948	3713	3414	2484
Total non-food crops	*15295*	*14326*	*13451*	*14183*	*8240*	*10345*	*8743*	*9031*	*10642*	*13096*	*10280*	*11029*
Total all crops	*48990*	*49289*	*49211*	*50461*	*46208*	*46341*	*49332*	*50028*	*55799*	*50666*	*48351*	*49012*
Net Area sown	*45509*	*45869*	*45539*	*45550*	*43237*	*44981*	*44886*	*42327*	*46208*	*44734*	*42628*	*42724*

Source: Taluka Panchayat Matar.

year-to-year, since 1967-68 (Table 2.15). Between 1963-64 and 1966-67 area under paddy varied between 11,533 and 13,463 hectares. The range has widened since then, area under paddy fluctuated after 1967-68 between 20,104 hectares (1971-72) and 13,235 hectares (1974-75). Area under wheat increased slowly but continuously from 7,500 hectares in 1963-64 to 9,229 hectares in 1974-75.

The wide fluctuations in area under paddy reflect largely fluctuations in total cropped area. Gross cropped area reached peak of 55,799 hectares in 1971-72, it declined to 48,351 hectares in 1973-74 and reached 49,012 hectares level in 1974-75. The wide fluctuations in gross cropped area are to be weighed against a declining trend in the net sown area. Fluctuations in net sown area were very small. With the exception of peak of 46,208 hectares in 1972-73 and trough of 42,327 hectares in 1970-71, the net sown area showed a decline from high 45,869 in 1964-65 to low of 42,724 hectares in 1974-75.

A decline in net sown area is compensated by increased gross cropped area suggesting an expansion of double cropped area. Irrigation helped in some parts expansion of area under second season crop, mainly wheat. Expansion of double cropped area was helped also by a decline in area under cotton crop which occupies land for nearly two seasons. Cotton has been replaced by a combination of paddy and wheat. Earlier in Matar Taluka, especially in Navagam region, long duration superior variety of paddy was grown. Introduction of H.Y.V. enabled to reduce time required for paddy and made possible growing of wheat as a second crop. This major change in cropping pattern is a direct result of a combination of increased supply of irrigation water and availability of short duration high yielding varieties of paddy and wheat. Unirrigated wheat was traditionally grown in Limbasi region. It is now grown over much wider area and as an irrigated crop.

Manures

There has been a qualitative shift in the use of manures, beside the increase in their quantity (Table 2.16). Ammonium sulphate has been used since 1963-64, its quantity fluctuated between 15,914 quintals (1965-66) and 5,600 quintals (1973-74). There was a sharp increase in the use of phosphates. Use of super phosphate increased from 145 quintals in 1963-64 to 1,900 quintals in 1974-75. Use of other fertilisers which included mainly urea, increased rapidly from 6,500 quintals in 1963-64 to 27,000 quintals in 1974-75. Urea thus acquired dominant position. Use of traditional manures, oil cakes and green manuring too increased fairly rapidly. Seeds of green manures distributed were 151 quintals in 1963-64 and 1,500 quintals in 1974-75.

TABLE 2.16: Distribution of Fertilisers and Manure

(In quintals)

Year	Ammonium Sulphate	Super Phosphate	Other Chemical Fertilisers	Oil Cake	Green Manure (Seeds)
1	2	3	4	5	6
1963-64	10000	145	6500	—	151
1964-65	13700	193	7300	—	386
1965-66	15914	183	7600	—	972
1966-67	14200	120	7450	—	850
1967-68	10160	2640	11700	—	650
1968-69	14950	1392	10690	3033	1727
1969-70	14000	703	14710	3670	1030
1970-71	11000	1000	15000	4000	1100
1971-72	9000	600	14000	4500	1100
1972-73	7000	500	16000	5100	1300
1973-74	5600	800	17000	5300	1200
1974-75	8100	1900	21000	5800	1500

Source: Office of the Taluka Development Officer, Matar. Panchayat Office.

Dairy Cattle, Bullocks and Tractors

A decline in number of bullocks and an addition to the number of buffaloes and a steep increase in the number of tractors in the last decade, suggest a shift in favour of single purpose animals and mechanisation of motive power. A possible decline in male cattle in future is suggested by a steeper decline in younger male cattle of ages three years and less.

Number of tractors increased from 47 in 1965-66 to 137 in 1972. The number of traditional ploughs registered a marginal increase from 7,273 to 7,865 between 1965 and 1972 (Table 2.17). If a tractor ploughs on an average 200 hectares of land during an agricultural year, 137 tractors would have potential to cover 27,400 hectares.

TABLE 2.17: Implements and Agricultural Machinery in Matar Taluka

Year	Ploughs		Carts	Oil Engines	Electric Motors	Tractors
	Wooden	Iron				
1	2	3	4	5	6	7
1965	7273	718	3843	262	77	47
1972	7865	722	4603	673	220	137

Source: Livestock Census, 1966 and 1972.

Expansion of dairy activity is suggested by two independent facts. Milk Cooperatives are an organisational arm of the indigenous industry, linked with a pioneering milk marketing institution of AMUL located at Anand in Kheda district. Milk cooperatives in Matar taluka increased from 32 to 62, between 1965-66 and 1974-75, the membership of these cooperatives increased from 5,722 to 11,371 or by about 93 per cent during the same period (Table 2.18). Total milk cooperative membership represented about 44 per cent of the total households in the taluka, they suggest a widespread coverage. In 1972, total 'animals in milk' were 12,175, of them 'buffaloes in milk' were 9,731 and 'Cows in milk' were 2,444. Thus buffaloes as a dairy animal predominate (Table 2.19).

TABLE 2.18: Milk Co-operatives in Matar Taluka

Year	Number of Milk Co-operatives	Number of Members
1	2	3
1965-66	32	5,722
1974-75	62	11,371

Source: Office of the Taluka Development Officer, Matar.

However, productivity of milk animals was low. Estimated milk production was 1,59,85,300 litres in 1974-75 which works out to 3.6 litres per animal in milk per day (Table 2.20). By any accepted standard this is a low level of milk yield. In milk production Matar and Limbasi regions were in lead with 37 and 35 per cent shares in total milk production in 1974-75. Share of Navagam was 28 per cent only.

TABLE 2.19: Bovine Population in Matar Taluka

(Number)

Bovine	1961	1966	1972
1	2	3	4
I *Cattle*			
A. Males over 3 years	14,331	14,805	12,785
B. Females over 3 years			
1. In Milk	—	—	2,444 (48.8)
2. Dry	—	—	2,348 (46.8)
3. Others	—	—	221 (4.4)
4. Total	5,369 (100)	5,550 (100)	5,013 (100)
C. Youngstock (under 3 years)	5,580	6,545	4,896
Cattle : Total :	25,280	26,900	22,694
II *Buffaloes*			
A. Males over 3 years	93	102	183
B. Females over 3 years			
1. In Milk	13,741	14,339	9,731 (57.2)
2. Dry	—	—	6,807 (40.0)
3. Others	21	83	469 (2.8)
4. Total	13,762	14,422	17,007 (100)
C. Youngstock (under 3 years)	8,345	11,308	13,666
Buffalo : Total :	22,200	25,832	30,856
III *Total Bovine*			
A. Male (over 3 years)	14,424	14,907	12,968
B. Female (over 3 years)	19,131	19,972	22,020
C. Youngstock (under 3 years)	13,925	17,853	18,562
D. Total	47,480	5[illegible],732	53,550

Source: Livestock Census, 1961, 1966, 1972.

TABLE 2.20: Estimated Milk Production in Matar Taluka (1974-75)

Region	Milk Production in '00 Litres	Percentage
1	2	3
Navagam	44,392	28
Matar	58,764	37
Limbasi	56,697	35
Total: Taluka	1,59,853	100

Source: Village Schedules, 1974-75.

In 1972, 40 and 46.8 per cent of buffaloes and cows respectively were dry. High percentage of buffaloes in milk and higher milk yield per buffalo compared to that for cow encouraged a shift in composition of bovine in favour of buffaloes. Between 1961 and 1972, number of adult cows declined from 5369, to 5013 i.e., by 8 per cent, number of adult buffaloes increased from 13,762 to 17,007 or by 24 per cent. A decline in youngstock of cattle from 5,580 in 1961 to 4,896 in 1972 and an increase in young stock of buffaloes from 8,345 in 1961 to 13,666 in 1972 emphasise a sharpers shift in favour of buffalo as a dairy animal in years following 1972. There was a decline in the importance of bullocks as a source of motive power in favour of mechanised motive power supplied by tractors. Together with tractors, tractor operated implements too have come in use. In 1972, 77 tractor cultivators, 30 levellers and 50 trailors were in use.

Summing-up

Matar of 1974-75 differs from that of 1965-66 in many vital respects. Three regions of the taluka had varying extent of area under irrigation in 1965-66, in 1974-75 they attained near equality in the extent of area irrigated. Nevertheless inter-regional differences persisted in a different sense. More rapid progress was witnessed in irrigation by Matar and Limbasi regions, but Navagam attained greater dependability. Limbasi experienced preponderance of canal water supply, cheaper but not as reliable as the other sources of irrigation. Limbasi lost larger area of land from under cultivation due to water-logging and water salinity. On the whole, use of fertilisers, more balanced ones, increased. Mechanisation and dairy development showed rapid expansion. The taluka as a whole witnessed wide fluctuations in area irrigated by different sources before 1967-68. After 1967-68 we find greater fluctuations in land allocated to different crops.

Significant regional differences, with increased overall irrigation and modernisation and expansion of dairy, an adjunct of crop-cultivation, provide the background, as does the sustained high rate (about 3 per cent simple rate of increase per year) of growth of population, for the analysis of the associated changes undertaken in the following chapters.

3

STRUCTURE OF LAND HOLDING

IN THE first resurvey we had observed enlargement of the class of wage-earners between 1929-30 and 1965-66. The observation was based on the comparison of two period data. For 1974-75 we have data for the same families at two points of time. These data bring out occupational movement of individual families over the nine-year period. The analysis brings out element of dynamism in the economy, the movements of the families studied were both upward and downward, and also in and out of cultivation. Such dynamism obtained despite increasing pressure of population on land. Rising incomes of families in different classes could be one of the factors generating such dynamism. Phenomenon of shrinking size of land available per capita is commonly observed in developing economies. Rapid increase in population and limited opportunities outside agriculture are two main factors responsible for this phenomenon.

The rising pressure of population on land may or may not be accompanied by reduced size of operational holding. If enough number of persons are cased out from cultivation of land and they join the ranks of landless labour, the size of operational holdings may remain unchanged. Instead, we may find size of operational holding declining where more people are absorbed in agriculture by way of subdivision of holdings.

Our observation for the period from 1929-30 to 1965-66 was that alternative job opportunities had declined but the operational holding size had not declined mainly because addition to rural population in the taluka swelled the ranks of landless labourers. Between 1965-66 and 1974-75, the rate of increase in population accelerated and per capita cultivable land tended to decline. We shall examines in this chapter, the changes in land distribution pattern between 1965-66 and 1974-75 both for the taluka as a whole and for the sample households.

We begin with changes in population and area of cultivable land in the taluka (Table 3.1).

TABLE 3.1: Population, Occupied Cultivable Land and Per Capita Land

Year	Rural Population (Number)	Percentage change during the interval	Compound rate of change per cent per annum	Occupied cultivable land (acres)	Percentage change during the interval	Compound rate of change per cent per annum	Land per capita (acres)	Compound rate of change percent per annum
1	2	3	4	5	6	7	8	9
1929-30*	58,965	—	—	95,737	—	—	1.65	—
1960-61*	1,01,319	+71.80	1.75	1,11,418	+16.40	+0.51	1.10	−1.32
1965-65*	1,16,514	+14.80	2.80	1,11,315	−0.09	N.S.	0.95	−2.90
1970-71**	1,32,084	+13.40	2.55	1,10,184	−1.00	−0.20	0.83	−2.67
1974-75†	1,52,212	+15.20	3.60	1,07,306	−2.60	−0.65	0.70	−4.17

N.S. = Not Significant

Note : Population level for 1965-66 and 1974-75 are estimated.

Sources : * Vimal Shah and C.H. Shah, *Resurvey of Matar Taluka*, 1974.

** Population Census, 1971.

† District Statistical Abstract (Kheda), 1978-79.

As already observed in Chapter 2 population growth has been relatively faster in Matar Taluka between 1951 and 1961 and between 1961 and 1971. The area under the occupied land during this period had started shrinking. The rate of decline accelerated during the two quinquennia, 1965-66 to 1970-71 and 1970-71 to 1974-75. Thus the pressure on land, though high in 1929-30, became more acute after 1951 and 1961. The average cultivable land per capita of rural population declined all through, from 1.65 in 1929-30 to 0.70 acres in 1974-75. Over 45 years, i.e., between 1929-30 and 1974-75, the average land per capita of rural population tended to decline. Between 1929-30 and 1960-61 it declined at the rate of 1.32 per cent. The rate of decline accelerated to 2.9 per cent per annum during 1960-61 and 1965-66 and thereafter it marginally decelerated to 2.67 per cent per annum between 1965-66 and 1970-71. Between 1970-71 and 1974-75 the per capita land availability declined at a high rate 4.2 per cent per annum (Table 3.1). Thus there was no letting in the tendency of mounting population continuing to press on land.

The rising pressure of population does not seem to have affected the average size of holding, a technical or an economic unit of operation, to the same extent, as the decline observed in regard to average land available per capita. Table 3.2 gives relevant data.

TABLE 3.2: Average Operational Holding in Matar Taluka

(Area in Acres)

Year	Number of holdings	Area of land held	Average size of holding	Percentage change
Holding in Taluka				
1929-30	19,273	95,737	4.97	—
1965-66	20,850	1,11,315	5.34	+7.44
1976-77	21,232	1,08,610	5.12	−4.11
Sample Holdings				
1965-66	628	5,647	8.99	—
1974-75	635	4,295	6.76	−24.8

Sources: 1. Vimal Shah and C.H. Shah, *Resurvey of Matar Taluka*, 1974.
2. Department of Agriculture, Government of Gujarat.
3. Field Survey of 1974-75.

Two contrasting trends were observed for the period 1929-30 and 1965-66 when we compare the changes in land per capita (a surrogate of population pressure) and the average size of holding. While the former declined by 42 per cent, the latter tended to increase by a small margin of 7 per cent. For all practical purposes the average size of holding over 36 years remained unchanged. The population increase, it would be inferred, swelled the ranks of landless labour. During 1965-66 and 1976-77, the average size of holding declined by 4.18 per cent, a much smaller decline compared to 26.32 per cent in the land per capita during the same period. Once again the phenomenon of population increase pressing more on the ranks of labourers is evident in the taluka.

The sample data on land holdings and their distribution offer scope for closer examination. In 1965-66, 628 holdings were reported. The average size of holding for the sample was 8.99 acres which was about 68 per cent larger than the average size of holding for the total taluka holdings computed on the basis of official records. This difference could be due to different reasons. One reason could be the land ceiling legislation. On record the holdings may have been sub-divided within the members of joint families but in reality the families may have continued to operate total land as one holding. Since the 1965-66 resurvey did not examine aspects relating to land records and land market, we are not in a position to offer adequate explanation for the difference observed in the average size of land holding for the sample and for the population. The average size of land holding for the 1974-75 sample worked out to be 6.76 acres which was 24.8 per cent less than the sample average size obtained for the 1965-66 sample.

Concentration of Land Holdings

An important question arises, pertaining to the distribution of land among different size groups. For examining the changing pattern of land distribution, we go back to 1929-30. Thus we shall capture a long term trend in distribution of holdings in the taluka as a whole. We repeat the exercise for the sample as well. Table 3.3 gives the data for the taluka.

The data in Table 3.3 show a decline in the inequality in land distribution over 1929-30 to 1976-77. The Gini Coefficient declined from 0.52 in 1929-30 to 0.43 in 1965-66, and for 1976-77 it was 0.44. In other words Gini coefficient declined during first 35 years but remained almost unchanged during the following decade. The average size of holding increased when the Gini ratio declined during 1929-30 and 1965-66, while average size declined during the following period when Gini ratio was unchanged.

TABLE 3.3: Distribution of Holdings and Area held in Matar Taluka

(Percentage)

Size of Holding (Acres)	Year					
	1929-30		1965-66		1976-77	
	Holdings	Area	Holdings	Area	Holdings	Area
1	2	3	4	5	6	7
0 to 5	75.0	29.3	71.3	32.1	65.2	26.9
5 to 15	19.2	31.5	22.8	40.5	25.9	39.5
15 to 25	3.5	13.1	4.7	16.6	7.5	24.6
25 and above	2.3	26.1	1.2	10.8	1.3	9.0
All Cultivators	100.0	100.0	100.0	100.0	100.0	100.0
Absolute values (No.& Acres)	19,273	95,737	20,850	1,11,315	21,232	1,08,610
Gini coefficient	0.52		0.43		0.44	

Sources: 1. Vimal Shah and C.H. Shah, *Resurvey of Matar Taluka*, Vora & Co., 1974.
2. Agricultural Census, 1976-77.

Gini ratio is a summary measure. A detailed examination of land distribution provides further insight into the changes. In 1929-30, 75 per cent of holdings were small (below 5 acres), in 1965-66, their percentage declined marginally to 71 per cent, but by 1976-77, decline was sharp. The share of small holdings was 65.2 per cent. The share of small holdings in total operated land also changed, it increased from 29.3 per cent in 1929-30 to 32.1 in 1965-66, but subsequently it declined to 26.9 per cent in 1976-77. The impact of these changes can be measured in terms of ratio of percentage share in land to percentage share in holdings. In the case of small holdings, the ratio moved from 0.39 to 0.45 in between 1929-30 and 1965-66, and then came down to 0.41 in 1976-77. Thus the relative position of small holdings first improved and then deteriorated. On the whole in a span of 45 years relative position of small holdings improved marginally. A similar exercise is carried out for other size groups and results are presented in Table 3.4.

It may be observed from Tables 3.3 and 3.4 that over time the proportion of holdings in the size groups of 5 to 15 acres increased. Their share in total land area increased substantially between 1929-30 and 1965-66, but it declined marginally between 1965-66 and 1976-77. Bigger size groups have lost in their relative importance all through.

TABLE 3.4: Ratio of Percentage of Land Area to Percentage of Holdings in Total in Matar Taluka

Size of Holdings (Acres)	1929-30	1965-66	1976-77
0 to 5	0.39	0.45	0.41
5 to 15	1.60	1.77	1.52
15 tc 25	3.74	3.53	3.28
25 and above	11.35	9.00	6.90

We now examine the Gini Coefficients computed for sample households. Between 1965-66 and 1974-75, Gini Coefficient for the distribution of sample holdings registered a marginal decline from 0.46 to 0.42 (Table 3.5). Since the nature of 1974-75 sample is non-random, the observed decline in Gini Coefficient warrants some explanation.

TABLE 3.5: Distribution of Holdings and Area held by Sample Farm Households in Matar Taluka

(Percentage)

Size of Holdings (Acres)	1965-66		1974-75	
	Holdings	Area	Holdings	Area
1	2	3	4	5
0-5	49.5	16.3	52.1	19.7
5-10	26.3	25.0	29.4	32.6
10-15	12.7	20.4	11.3	20.6
15-25	6.5	14.8	3.6	9.6
25 +	4.9	23.5	3.5	17.5
All Holdings	100.0	100.0	100.0	100.0
Absolute Values	628	5647	635	4295
Gini Coefficient	0.46		0.42	

Sources: Field Survey, 1965-66 and 1974-75.

Between 1965-66 and 1974-75, there was some mobility among the occupation groups. For a few cultivator households main source of income changed. Similarly, for a few non-cultivator households too the main source of income changed in favour of cultivation. It becomes necessary therefore to analyse these changes. Table 3.6 and onwards contain the composition of cultivator households of 1965-66 resurveyed during 1974-75 and the composition of all households of 1974-75 sample.

TABLE 3.6: Composition of Sample Households in 1974-75 by Occupation (1965-66 Households that did not Branch Out)

Occupation in 1974-75	Occupation in 1965-66												
	Cultivators					All culti-vators	Agri-cultural labour-ers	Artisan	Trader	Service	Others	All Non-farms	All House-holds
	0-5	5-10	10-15	15-25	25 +								
Cultivators													
0-5 acres	121	21	8	—	1	151	25	2	—	1	2	5	181
5-10 acres	27	64	12	1	—	104	6	—	1	1	2	4	114
10-15 acres	5	20	30	3	—	58	—	—	—	—	1	1	59
15-25 acres	5	10	3	10	1	29	—	—	—	—	—	—	29
25 +	2	1	4	3	12	22	—	—	—	—	—	—	22
All Cultivators	160	116	57	17	14	364	31	2	1	2	5	10	405
Agricultural Labourers	15	2	—	—	—	17	157	2	1	3	9	15	189
Artisan	4	1	—	—	—	5	—	22	—	—	3	25	30
Trader	2	—	—	—	—	2	—	—	26	1	3	30	32
Service	8	1	2	—	1	12	12	—	3	22	2	27	51
Others	4	—	1	1	—	6	3	3	1	1	25	30	39
All Non-farms	18	2	3	1	1	25	15	25	30	24	33	112	152
All Households	193	120	60	18	15	406	203	29	32	29	47	137	746

TABLE 3.7: Composition of Sample Households in 1974-75 by Occupation (1965-66 Households Reporting Satellites in 1974-75)

Occupation in 1974-75	Occupation in 1965-66												
	Cultivators					All culti-vators	Agri-cultural labour-ers	Artisan	Trader	Service	Others	All Non-farms	All House-holds
	0-5	5-10	10-15	15-25	25 +								
Cultivators													
0-5	13	8	—	—	—	21	3	—	—	—	—	—	24
5-10	10	11	2	—	—	23	1	—	—	—	—	—	24
10-15	4	6	4	—	—	14	—	—	—	—	—	—	14
15-25	1	2	2	2	1	8	—	—	—	—	—	—	8
25 +	—	1	1	3	5	10	—	—	—	—	—	—	10
All Cultivators	28	28	9	5	6	76	4	—	—	—	—	—	80
Agricultural Labourers	2	—	—	—	—	2	18	—	—	—	—	—	20
Artisan	—	—	—	—	—	—	—	2	—	—	—	2	2
Trader	1	—	—	—	—	1	—	—	—	—	—	—	1
Service	1	1	—	—	—	2	1	—	—	—	—	—	3
Others	—	1	—	—	—	1	1	—	—	—	—	—	2
All Non-farms	2	2	—	—	—	4	2	2	—	—	6	8	14
All Households	32	30	9	5	6	82	24	2	—	—	6	8	114

TABLE 3.8: Composition of Sample Households in 1974-75 by Occupation (Satellite Households)

Occupation in 1974-75	Occupation in 1965-66												
	Cultivators					All cultivators	Agri-cultural labourers	Artisan	Trader	Service	Others	All Non-farms	All House-holds
	0-5	5-10	10-15	15-25	25 +								
Cultivators													
0-5	9	5	—	—	—	14	20	—	1	2	—	3	37
5-10	17	6	—	—	—	23	6	—	3	4	1	8	37
10-15	6	8	2	—	—	16	5	—	1	2	—	3	24
15-25	3	2	4	1	5	15	4	—	1	—	—	1	20
25 +	6	5	1	—	3	15	1	—	—	—	—	—	16
All Cultivators	41	26	7	1	8	83	36	—	6	8	1	15	134
Agricultural Labourers	—	—	—	—	—	—	25	—	1	3	1	5	30
Artisan	—	—	—	—	—	—	—	3	—	—	—	3	3
Traders	—	—	—	—	—	—	1	—	—	—	—	—	1
Service	1	—	—	—	—	1	3	—	1	3	—	4	8
Others	1	—	—	—	—	1	3	—	—	—	7	7	11
All Non-farms	2	—	—	—	—	2	7	3	1	3	7	14	23
All Households	43	26	7	1	8	85	68	3	8	14	9	34	187

We have presented tables in two formats. In one set of tables (Tables 3.6 to 3.8) we have presented composition of sample households not reporting any branching out during 1965-66 and 1974-75, sample households that reported branching out and the satellite households that did not exist during 1965-66 but came into existence in the intervening period. The ten tables given in Appendix 3.1 to 3.10 give a detailed composition of households in 1974-75 for ten major occupations.

Of the 364 cultivating households not experiencing branching out and remaining in the occupation of cultivation both during 1965-66 and 1974-75, 237 or about 70 per cent continued with same size of operational holding during two survey years. Nearly 30 per cent of households experienced change in the size of operational holding. Of 127 households experiencing change in the size of operational holding, 80 or about 65 percentage experienced a decline in their operational land holding in 1974-75. Thus, of the households reporting operational holding of 0-5 and 5-10 acres in 1974-75, a large number of them came from operational holding size of 10 acres and above in 1965-66.

Relatively fewer households – about 10 per cent, reporting cultivation as their occupation in 1974-75 belonged to either agricultural labourer's class or to other non-farm activities during 1965-66. Of the 17 agricultural labourer households which joined cultivation group in 1974-75 and 15 did operate land even in 1965-66. Similarly, 20 out of 25 non-farm households in 1965-66 reporting cultivation as their occupation in 1974-75 did operate land even in 1965-66. What is interesting to note for agricultural labourers and other non-farm households is that the average size of operational holding in 1974-75 was generally higher than that of the size reported during 1965-66. Since, hardly any sales transactions were reported in the intervening period, it can be said that many cultivating households might have leased out the land to people engaged as agricultural labourers or other non-farm activities during 1965-66.

The number of agricultural labour households not reporting any branching out increased from 189 to 203 between 1965-66 and 1974-75. Thirty two households out of 189 in 1965-66 reported a change in their main source of livelihood and 47 households out of 203 in 1974-75 joined the class of agricultural labourers, coming mainly from cultivating classes.

The occupational mobility among the cultivating households reporting branching out between 1965-66 and 1974-75 was relatively higher. Forty one out of 76 cultivating households belonged to a different size-class of operational holding than the one which was reported during 1965-66. Thirty out of 41 households reporting change in the size of operational holding experienced a reduction in size of their land holding.

From among all other occupations the mobility was limited. The composition of satellite households based on their occupation categories in 1965-66 and the occupation of their parent households suggests high mobility. Nearly 80 per cent of the households reporting operational holding of 0.5 acres in 1974-75 belonged to parent households who had reported more than 5 acres of operational holding during 1965-66. Similarly, 36 or about 25 per cent of households, whose parent households belonged to cultivation class, joined the agricultural labourer class.

In all changes in the composition of two classes are worth observing, namely, 0-5 acres size group and the agricultural labourer class. Of the 268 total households reporting their occupation in the 0-5 acres class of land holding category (excluding new households surveyed during 1974-75) 143 or about 44 per cent had a different occupation in 1965-66. Of these 143, 86 households belonged to cultivating households reporting holdings bigger than 5 acres in 1965-66.

Similarly, out of 296 agricultural labourer households in 1974-75 (excluding 18 new households) 95 households belonged to different occupation groups in 1965-66. Out of 95, 71 belonged to cultivating households (Appendix Table 3.6). Thus for a substantial number of households the mobility experienced between 1965-66 and 1974-75 meant joining economically lower ranking occupation group of course with probably a better base than their other counterparts in the occupation group during 1974-75.

Before the discussion on the land size and its related issue is closed, it is worth observing that the average size of holding reduced substantially for the 0-5, 5-10 and 15-25 acres size group between 1965-66 and 1974-75. For 10-15 acres size group the decline in average size of holding was marginal and for 25+ acres size group, it was negligible. General diminution of size of holding could be due to two factors, bequethal on death of the head of the family or on division of the household during the lifetime of the head. It could also be due to transactions in land market through sale, mortgage, and lease. Analysis of changes in the size of holdings without satellite families, as apart from others, enables us to draw an inference that, beside the surge in population, active land market too had a role in affecting the size of holding. The exact magnitude and the nature of land market transactions could not be determined in absence of requisite data. We made an attempt to obtain the data during the field survey, the reported transactions are insignificant, suggesting sizeable underreporting of active land market.

APPENDIX 3.1

Composition of Sample Households With an Operational Holding of 0-5 Acres/Occupation in 1974-75 in Matar Taluka

Occupation	No. of HHs without Satellite	Average size of holding, 1965-66	Average size of holding, 1974-75	No. of HHs with Satellite	Average size of holding, 1965-66	Average size of holding, 1974-75	No. of Satellite HHs with land	Average size of holding, 1974-75	No. of Satellite HHs without land
1	2	3	4	5	6	7	8	9	10
0—5	121	3.22	2.59	13	4.07	3.37	3	1.52	6
5—10	27	8.03	3.03	10	9.20	3.32	8	2.91	9
10—15	5	14.60	3.94	4	20.25	3.19	2	2.96	4
15—25	5	23.00	3.14	1	24.00	1.98	3	3.11	—
25 +	2	47.00	4.30	—	—	—	5	4.08	1
Cultivators	160	5.55	2.75	28	8.92	3.28	21	3.02	20

Contd.

Contd.

1	2	3	4	5	6	7	8	9	10
Agri. Labour	15	2.20[1]	2.60	2	2.00[2]	2.80[11]	—	—	—
Artisan	4	4.00	2.75	—	—	—	—	—	—
Traders	2	0.00	2.56	1	5.00	3.09	—	—	—
Service	8	2.39[3]	2.10	1	4.00	3.00	1	1.20	—
Others	4	4.00[4]	2.76[5]	-	—	—	1	2.00	—
Non-Farm	18	3.06[6]	2.37[7]	2	4.5	3.04	2	1.60	—
All HHs	193	5.14[8]	2.71[9]	32	8.41[10]	3.23	23	2.90	20

1 HHs reporting land 10
2 HHs reporting land 1
3 HHs reporting land 7
4 HHs reporting land 1
5 HHs reporting land 2
6 HHs reporting land 12
7 HHs reporting land 18
8 HHs reporting land 222
9 HHs reporting land 232
10 HHs reporting land 31
11 HHs reporting land 2

4 NEW FAMILIES

APPENDIX 3.2

Composition of Sample Households With an Operational Holding of 5-10 Acres/Occupation in 1974-75 in Matar Taluka

Occupation	No. of HHs. without Satellite	Average size of holding, 1965-66	Average size of holding, 1974-75	No. of HHs. with Satellite	Average size of holding, 1965-66	Average size of holding, 1974-75	No. of Satellite HHs with land	Average size of holding, 1974-75	No. of Satellite HHs without land
1	2	3	4	5	6	7	8	9	10
0 — 5	21	3.80	6.80	8	4.38	6.60	—	—	5
5 — 10	64	8.66	7.41	11	8.55	7.37	—	—	6
10 — 15	20	12.50	8.60	6	17.00	7.20	4	7.22	4
15 — 25	10	25.10	8.56	2	17.00	8.77	2	7.79	—
25 +	1	52.00	7.35	1	38.00	9.86	4	7.02	1
Cultivators	116	10.23	7.60	28	10.82	7.31	10	7.25	16

Contd.

Contd.

1	2	3	4	5	6	7	8	9	10
Agri. Labour	2	—	6.56	—	—	—	—	—	—
Artisan	1	5.00	5.24	—	—	—	—	—	—
Traders	—	—	—	—	—	—	—	—	—
Service	1	—	6.28	1	7.00	7.51	—	—	—
Others	—	—	—	1	2.00	8.00	—	—	—
Non-farm	2	5.00*	5.76	2	4.50	7.76	—	—	—
All HHs	120	10.18**	7.55	30	10.40	7.34	10	7.25	16

* HHs reporting land 1

** HHs reporting land 117

1 NEW FAMILY

APPENDIX 3.3

Composition of Sample Households With an Operational Holding of 10-15 Acres/Occupation in 1974-75 in Matar Taluka

Occupation	No. of HHs. without Satellite	Average size of holding, 1965-66	Average size of holding, 1974-75	No. of HHs. with Satellite	Average size of holding, 1965-66	Average size of holding, 1974-75	No. of Satellite HHs with land	Average size of holding, 1974-75	No. of Satellite HHs without land
1	2	3	4	5	6	7	8	9	10
0 – 5	8	4.13	12.08	—	—	—	—	—	—
5 – 10	12	5.42	4.60	2	14.00	11.46	—	—	—
10 – 15	30	13.93	11.89	4	18.25	12.40	—	—	2
15 – 25	3	20.67	12.25	2	17.00	12.61	—	—	4
25 +	4	40.00	15.24	1	47.00	14.55	1	14.13	—
Cultivators	57	12.95	12.11	9	20.33	12.47	1	14.13	6

Contd.

Contd.

1	2	3	4	5	6	7	8	9	10
Agri. Labour	—	—	—	—	—	—	—	—	—
Artisan	—	—	—	—	—	—	—	—	—
Traders	—	—	—	—	—	—	—	—	—
Service	2	12.00	14.82	—	—	—	—	—	—
Others	1	—	10.35	—	—	—	—	—	—
Non-farm	3	12.00*	13.33	—	—	—	—	—	—
All HHs	60	12.92**	12.18	9	20.33	12.47	1	14.13	6

* HHs reporting land 2
** HHs reporting land 59

APPENDIX 3.4

Composition of Sample Households With an Operational Holding of 15-25 Acres/Occupation in 1974-75 in Matar Taluka

Occupation	No. of HHs without Satellite	Average size of holding, 1965-66	Average size of holding, 1974-75	No. of HHs with Satellite	Average size of holding, 1965-66	Average size of holding, 1974-75	No. of Satellite HHs with land	Average size of holding, 1974-75	No. of Satellite HHs without land
1	2	3	4	5	6	7	8	9	10
0 – 5	—	—	—	—	—	—	—	—	—
5 – 10	1	7.00	19.21	—	—	—	—	—	—
10 – 15	3	16.00	16.55	—	—	—	—	—	—
15 – 25	10	17.00	16.08	2	25.50	23.39	—	—	1
25 +	3	47.67	18.99	3	35.67	19.27	—	—	—
Cultivators	17	21.65	16.66	5	31.60	20.92	—	—	1

Contd.

Contd.

1	2	3	4	5	6	7	8	9	10
Agri. Labour	—	—	—	—	—	—	—	—	—
Artisan	—	—	—	—	—	—	—	—	—
Traders	—	—	—	—	—	—	—	—	—
Service	—	—	—	—	—	—	—	—	—
Others	1	14.00	22.23	—	—	—	—	—	—
Non-farm	1	14.00	22.23	—	—	—	—	—	—
All HHs	18	21.23	16.97	5	31.60	20.92	—	—	1

APPENDIX 3.5

Composition of Sample Households With an Operational Holding of 25 + Acres/Occupation in 1974-75 in Matar Taluka

Occupation	No. of HHs without Satellite	Average size of holding, 1965-66	Average size of holding, 1974-75	No. of HHs with Satellite	Average size of holding, 1965-66	Average size of holding, 1974-75	No. of Satellite HHs with land	Average size of holding, 1974-75	No. of Satellite HHs without land
1	2	3	4	5	6	7	8	9	10
0 – 5	1	3.00	27.67	—	—	—	—	—	—
5 – 10	—	—	—	—	—	—	—	—	—
10 – 15	—	—	—	—	—	—	—	—	—
15 – 25	1	25.00	34.26	1	17.00	37.94	—	—	5
25 +	12	41.42	32.31	[illegible]	46.60	34.04	3	32.77	—
Cultivators	14	37.50	37.47	6	41.17	34.69	3	32.77	5

Contd.

Contd.

1	2	3	4	5	6	7	8	9	10
Agri. Labour	—	—	—	—	—	—	—	—	—
Artisan	—	—	—	—	—	—	—	—	—
Traders	—	—	—	—	—	—	—	—	—
Service	1	—	55.84	—	—	—	—	—	—
Others	—	—	—	—	—	—	—	—	—
Non-farm	1	—	55.84	—	—	—	—	—	—
All HHs	15	37.50	38.85	6	41.17	34.69	3	32.77	5

APPENDIX 3.6

Composition of Sample Households With an Occupation/Agricultural Labour in 1974-75 in Matar Taluka

Occupation	No. of HHs without Satellite	Average size of holding, 1965-66	Average size of holding, 1974-75	No. of HHs with Satellite	Average size of holding, 1965-66	Average size of holding, 1974-75	No. of Satellite HHs with land	Average size of holding, 1974-75	No. of Satellite HHs without land
1	2	3	4	5	6	7	8	9	10
0 – 5	25	3.44	1.62[1]	3	2.67	1.83[2]	—	—	20
5 – 10	6	7.50	3.87[3]	1	6.00	5.98	—	—	6
10 – 15	—	—	—	—	—	—	—	—	5
15 – 25	—	—	—	—	—	—	—	—	4
25 +	—	—	—	—	—	—	—	—	1
Cultivators	31	4.23	2.03[4]	4	3.50	3.21	—	—	36

Contd.

Contd.

1	2	3	4	5	6	7	8	9	10
Agri. Labour	157	1.61[5]	1.68[6]	18	2.40[7]	2.09[8]	1	0.63	25
Artisan	—	—	—	—	—	—	—	—	—
Traders	—	—	—	—	—	—	—	—	—
Service	12	1.00[9]	—	1	1.00	0.56	—	—	3
Others	3	—	2.75[10]	1	—	2.33	—	—	3
Non-farm	15	1.00[9]	2.75[10]	2	1.00[11]	1.43	—	—	7
All HHs	203	2.92[12]	1.87[13]	24	2.60[14]	2.27[15]	1	0.63	68

1 HHs reporting land 9
2 HHs reporting land 2
3 HHs reporting land 2
4 HHs reporting land 11
5 HHs reporting land 28
6 HHs reporting land 19
7 HHs reporting land 10
8 HHs reporting land 12
9 HHs reporting land 2
10 HHs reporting land 2
11 HHs reporting land 1
12 HHs reporting land 61
13 HHs reporting land 32
14 HHs reporting land 15
15 HHs reporting land 18

18 NEW FAMILIES

APPENDIX 3.7

Composition of Sample Households With an Occupation/Artisan in 1974-75 in Matar Taluka

Occupation	No. of HHs without Satellite	Average size of holding, 1965-66	Average size of holding, 1974-75	No. of HHs with Satellite	Average size of holding, 1965-66	Average size of holding, 1974-75	No. of Satellite HHs with land	Average size of holding, 1974-75	No. of Satellite HHs without land
1	2	3	4	5	6	7	8	9	10
0 — 5	2	2.50	5.30[1]	—	—	—	—	—	—
5 — 10	—	—	—	—	—	—	—	—	—
10 — 15	—	—	—	—	—	—	—	—	—
15 — 25	—	—	—	—	—	—	—	—	—
25 +	—	—	—	—	—	—	—	—	—
Cultivators	2	2.50	5.30[1]	—	—	—	—	—	—

Contd.

Contd.

1	2	3	4	5	6	7	8	9	10
Agri. Labour	2	—	0.90[2]	—	—	—	—	—	—
Artisan	22	2.46[3]	2.21[3]	2	6.00[4]	8.14[4]	—	—	3
Traders	—	—	—	—	—	—	—	—	—
Service	—	—	—	—	—	—	—	—	—
Others	3	—	—	—	—	—	—	—	—
Non-farm	25	2.46[3]	2.21[3]	2	6.00[4]	8.14[4]	—	—	3
All HHs	29	2.47[5]	1.87[5]	2	6.00[4]	8.14[4]	—	—	3

1 HHs reporting land 1
2 HHs reporting land 1
3 HHs reporting land 11
4 HHs reporting land 1
5 HHs reporting land 13

4 NEW FAMILIES

APPENDIX 3.8

Composition of Sample Households With an Occupation/Trader in 1974-75 in Matar Taluka

Occupation	No. of HHs without Satellite	Average size of holding, 1965-66	Average size of holding, 1974-75	No. of HHs with Satellite	Average size of holding, 1965-66	Average size of holding, 1974-75	No. of Satellite HHs with land	Average size of holding, 1974-75	No. of Satellite HHs without land
1	2	3	4	5	6	7	8	9	10
0—5	—	—	—	—	—	—	—	—	1
5—10	1	5.00	—	—	—	—	—	—	3
10—15	—	—	—	—	—	—	—	—	1
15—25	—	—	—	—	—	—	—	—	1
25 +	—	—	—	—	—	—	—	—	—
Cultivators	1	5.00	—	—	—	—	—	—	6

Contd.

Contd.

1	2	3	4	5	6	7	8	9	10
Agri. Labour	1	—	—	—	—	—	—	—	1
Artisan	—	—	—	—	—	—	—	—	—
Trader	26	3.84[1]	1.54[2]	—	—	—	—	—	—
Service	3	2.00[3]	1.03[4]	—	—	—	—	—	1
Others	1	—	—	—	—	—	—	—	—
Non-Farm	30			—	—	—	—	—	1
All HHs	32			—	—	—	—	—	8

1 HHs reporting land 6
2 HHs reporting land 4
3 HHs reporting land 1
4 HHs reporting land 2

3 NEW FAMILIES

APPENDIX 3.9

Composition of Sample Households With an Occupation/Service in 1974-75 in Matar Taluka

Occupation	No. of HHs without Satellite	Average size of holding, 1965-66	Average size of holding, 1974-75	No. of HHs with Satellite	Average size of holding, 1965-66	Average size of holding, 1974-75	No. of Satellite HHs with land	Average size of holding, 1974-75	No. of Satellite HHs without land
1	2	3	4	5	6	7	8	9	10
0 – 5	1	1.00	2.83	—	—	—	—	—	2
5 – 10	1	5.00	2.30	—	—	—	—	—	4
10 – 15	—	—	—	—	—	—	2	3.30	—
15 – 25	—	—	—	—	—	—	—	—	—
25 +	—	—	—	—	—	—	—	—	—
Cultivators	2			—	—	—	2		6

Contd.

Contd.

1	2	3	4	5	6	7	8	9	10
Agri. Labour	3	1.00[1]	0.60[1]	—	—	—	—	—	3
Artisan	—	—	—	—	—	—	—	—	—
Trader	1	—	—	—	—	—	—	—	—
Service	22	7.75[2]	5.95[2]	3	—	1.47[3]	—	—	3
Others	1	2.00	1.70	—	—	—	—	—	—
Non-farm	24			3	—	—	—	—	3
All HHs	29			3			2		12

1 HHs reporting land 1
2 HHs reporting land 4
3 HHs reporting land 2

29 NEW FAMILIES

APPENDIX 3.10

Composition of Sample Households With an Occupation/Others in 1974-75 in Matar Taluka

Occupation	No. of HHs without Satellite	Average size of holding, 1965-66	Average size of holding, 1974-75	No. of HHs with Satellite	Average size of holding, 1965-66	Average size of holding, 1974-75	No. of Satellite HHs with land	Average size of holding, 1974-75	No. of Satellite HHs without land
1	2	3	4	5	6	7	8	9	10
0 – 5	2								—
5 – 10	2								1
10 – 15	1								—
15 – 25									—
25 +									—
Cultivators	5								1

Contd.

Contd.

1	2	3	4	5	6	7	8	9	10
Agri. Labour	9								1
Artisan	3								
Trader	3								
Service	2								
Others	25			6			2		5
Non-farm	33			6			2		5
All HHs	47			6			2		7

1 NEW FAMILY

4

IRRIGATION: Sources, Use and Pattern

THE EFFECT of irrigation on any of the components of production is a part of the overall process of water use and water management. A change from extensive use of irrigation to intensive use of irrigation, from single source provided by public sector to conjunctive use of canal and other sources, are illustrations of improved water use. We investigate in this chapter, whether extension of irrigation is accompanied by improved water use and water management. We discuss from this angle, the intensity, conjunctive use, frequency of watering and extent of irrigation. We compare the change-over time regarding the extent of irrigation. Other aspects are discussed for 1974-75 only. We analyse irrigation data in detail for irrigation in Matar Taluka as already noted is the leading growth promoting factor.

Extent of Irrigation

Irrespective of the quantity and nature of water used any cropped area which receives water other than rain water is recorded as area under irrigation in the official records. This is not a happy description, for water is also supplied to crops to supplement rain water, when the latter is marginally deficient and can be made up by one or two waterings. Crops grown outside monsoon season need frequent and intensive use of water. To put on par area under irrigation in *Kharif* and in *Rabi*, is to ignore the function of water supplied at cost as an input for corp output. Despite this deficiency the measure of extent of area under irrigation is used, for it gives atleast an idea of the access to irrigation. Interpreted thus, the data regarding the extent of irrigation gives an extent of protection provided by irrigation water, in addition to rain water in *Kharif*.

The taluka level data regarding the extent of irrigation were examined earlier in Chapter 2. They revealed wide fluctuations from year-to-year even during the recent period when canal irrigation rapidly expanded. We have also seen the uneven spread of irrigation in different regions in 1965-66 and the progress of irrigation during 1965-66 and 1974-75 which narrowed the wide regional differences. In 1974-75, the cultivator households in the sample, had about 67 per cent of cropped area under irrigation. During the period from 1965-66 to 1974-75 when area under irrigation increased from 35 to 67 per cent, for the sample, increased irrigation made possible for cultivating households in different size groups to have a fairly comparable extent of area under irrigation (Table 4.1). If we excluded the group of large cultivators (25 acres and above), the extent of irrigation in 1974-75 varied, between 63 and 67 per cent for different size classes, the lowest being in the middle size class (10-15 acres). For large cultivators the extent of cropped area under irrigation was nearly 77 per cent.* In 1965-66, the extent of irrigation varied between 30 and 41 per cent, the lowest percentage obtained for 0-5 acres size class, but the highest obtained for middle group of 10-15 acres size class. There were no major inter-size variations in the proportions of cropped area under irrigation in either of the two years. Thus, on the whole, benefit of expansion of area under irrigation reached uniformly to all size classes of cultivators.

According to the sample household data, of the total cropped area, over two-thirds was irrigated. In the event of deficient rainfall, second crop is less probable. Area sown, i.e., net sown area, area on which some crop, first or second, is grown, matters. Extent of irrigation protection given to it ensures survival of the crop. Viewed thus 70 per cent of area under cultivation had access to irrigation. The small and the large cultivators (0 to 5 acres and 25 acres and above) had higher access to irrigation – 74 and 78 per cent respectively (Table 4.2). The lowest percentage was 63, for middle size group (10-15 acres). Navagam region had 84 per cent of area with an access to irrigation (Table 4.3), as against 73 per cent of cropped area under irrigation. In the other two regions, area with an access to irrigation and cropped area irrigated were 66 and 69 (Matar) and 64 and 58 per cent (Limbasi). These data suggest nearly full utilisation of irrigation sources.

---

* This result is in contrast with the data based on Land Census and discussed in chapter two could there be under reporting? We have no firm information.

TABLE 4.1: Extent of Irrigation by Size of Holding and Region-Area Under Irrigation as per cent of net Sown Area

Size of Holding (acres)	Years	Region			Total
		Navagam	Matar	Limbasi	
1	2	3	4	5	6
0 – 5	1965	46.9	22.0	22.2	29.8
	1975	64.7	69.6	54.2	64.8
5 – 10	1965	50.4	23.2	21.7	33.1
	1975	73.3	69.3	54.2	66.1
10 – 15	1965	54.3	28.9	30.9	40.6
	1975	67.5	65.3	52.5	62.5
15 – 25	1965	59.6	19.4	25.7	39.2
	1975	80.6	85.6	44.6	66.7
25 +	1965	55.0	35.5	19.4	32.1
	1975	100.0	42.7	72.0	76.7
Total	1965	52.4	25.2	22.3	34.9
	1975	73.5	68.7	58.4	66.7

Source: Field Survey, 1965-66 and 1974-75.

TABLE 4.2: Area with access to Irrigation Among Sample Households in Matar Taluka in 1974-75

Size of Holding (acres)	Total Operated area (Acres)	Total area under Cultivation (acres)	Area with access to irrigation (acres)	4÷3 x 100
1	2	3	4	5
0 – 5	692.6	671.6	498.9	74.3
5 – 10	1283.0	1242.0	864.4	69.6
10 – 15	857.2	826.1	521.5	63.1
15 – 25	420.4	402.3	281.3	69.9
25 and above	801.4	729.0	564.8	77.5
All Cultivators	4054.6	3871.0	2730.9	70.5

Source: Field Survey, 1974-75.

Access to irrigation may be alternatively viewed in terms of the percentage of holdings that received irrigation, as well as in terms of whether holdings concerned were fully or partly irrigated and if partly irrigated whether a small or major part of it was irrigated. Against 70 per cent of land with an access to irrigation over 88 per cent of land holdings had an access to irrigation (Table 4.4). About 86 per cent of

TABLE 4.3: Area with Access to Irrigation Among Sample Households in Different Regions of Matar Taluka in 1974-75

Region	Total Operated area (acres)	Total Cultivated Area (acres)	Area with access to irrigation (acres)	4÷3 × 100
1	2	3	4	5
Navagam	1224.2	1142.5	959.8	84.0
Matar	1359.2	1293.3	850.7	65.8
Limbasi	1471.2	1435.2	920.3	64.1
Matar Taluka	4054.6	3871.0	2730.8	70.5

Source: Field Survey, 1974-75.

TABLE 4.4: Holdings with Irrigation Facility in Sample Households in Matar Taluka, 1974-75

Size of Holding (acres)	Total Holdings	Holding with Irrigation Facility	3 ÷ 2 × 100
1	2	3	4
0 – 5	251	215	85.7
5 – 10	156	145	92.9
10 – 15	71	61	85.9
15 – 25	23	21	91.3
25 +	22	20	90.9
All Cultivators	523	462	88.3

Source: Field Survey, 1974-75.

TABLE 4.5: Holdings with Irrigation Facility in Sample Households in Different Regions of Matar Taluka, 1974-75

Size of Holding Acres	Total Holdings			Holdings with Irrigation Facility			Percentage of Holdings with irrigation facility		
	Navagam	Matar	Limbasi	Navagam	Matar	Limbasi	Navagam	Matar	Limbasi
1	2	3	4	5	6	7	8	9	10
0 to 5	65	136	50	60	112	43	92.3	82.3	86.0
5 to 10	52	60	44	49	54	42	92.2	90.0	95.5
10 to 15	26	26	17	20	25	16	76.9	89.3	94.1
15 to 25	7	7	9	6	6	9	85.7	85.7	100.0
25 and above	7	2	13	6	2	12	85.7	100.0	92.3
All Cultivators	157	233	133	141	199	122	89.8	85.4	91.7

Source: Field Survey, 1974-75.

small holdings (0-5 acres) had an access to irrigation. By any token, irrigation in Matar Taluka is widespread, almost all classes of cultivators had fairly sizeable area under irrigation. The regional differences were very small (Table 4.5).

The real differences among holdings arise regarding the degree of protection. Small farmers had the highest percentage, 75 per cent of holdings were fully irrigated. The percentage of fully protected holdings declined as the size of holding increased (Table 4.6).

TABLE 4.6: Size Groupwise Distribution of Irrigated Holdings, 1974-75

(per cent)

Size of Holding (acres)	Fully Irrigated (90% or more)	Partly Irrigated holding with			Total Irrigated Holdings
		89% to 50%	49% to 25%	Less than 25%	
1	2	3	4	5	6
0 – 5	74.9	12.8	9.6	2.7	100.0
5 – 10	50.9	26.4	15.7	6.9	100.0
10 – 15	43.6	27.4	16.1	12.9	100.0
15 – 25	42.9	33.3	14.3	9.5	100.0
25 +	36.4	50.0	13.6	-	100.0
Total	59.8	21.7	12.8	5.6	100.0

Source: Field Survey, 1974-75.

Inter-regional differences regarding the extent of irrigation protection to holdings are rather wide, fully protected holdings accounted for 84 per cent of the irrigated holdings in Navagam, and for 59 per cent in Matar and only 36 per cent in Limbasi. Though the overall per cent of irrigated holdings to total holdings varied between 85 (Matar) and 92 (Limbasi) per cent, wide variation in the extent of protection made vital difference to the type of irrigation and water management (Table 4.7).

Sources of Irrigation

Data regarding sources of irrigation refer to allocation of irrigated area to individual sources. We shall first review the taluka level position and then look at the sample data. The 1965-66 resurvey did not collect data on this aspect and hence only the 1974-75 sample data will be reviewed.

TABLE 4.7: Distribution of Irrigated Holdings (Regionwise)

Region	Fully Irrigated (90% and more)	Partly Irrigated holdings with			Total Irrigated Holdings
		89% to 50%	49% to 25%	Less than 25%	
1	2	3	4	5	6
Navagam	84.4	10.6	2.1	2.8	100.0
Matar	59.0	20.5	13.2	7.3	100.0
Limbasi	35.8	35.0	23.4	5.8	100.0
Total	59.8	21.7	12.8	5.6	100.0

Source: Field Survey, 1974-75.

The taluka level data on sources of irrigation are already given in Chapter 2, Table 2.15. We present in Table 4.8 the percentage distribution of area irrigated by source for the years 1965-66 and 1974-75 for Matar Taluka.

TABLE 4.8: Area Irrigated by Sources in Matar Taluka

(Per cent)

Year	Sources			Other Sources	Total
	Government Canals	Tanks	Wells		
1	2	3	4	5	6
1965-66	70.9	1.9	13.9	13.3	100.0
1974-75	70.3	4.5	24.4	0.8	100.0

Source: Taluka Development Officer, Matar.

It may be observed that the canals had an important share in the total irrigated area even in 1965-66. In both the years canals irrigated 70 per cent of the total cropped area. In 1974-75, wells became second most important source.

The taluka level data do not give any idea regarding the multiple use or conjunctive use of sources of irrigation which indicates intensive use and better management of available irrigation sources.

Sample data for 1974-75 enable us to trace the extent of conjunctive use of different sources of irrigation. Data are limited.

They however provide an important insight. In 1974-75 canals alone irrigated 52.26 per cent of the total area under irrigation, but 24.4 per cent of cropped area was irrigated by more than one source (Table 4.9). It was reported that the canals were one of the sources for all of them and hence total area irrigated by canals would amount to 77 per cent. These data suggest that nearly one-fourth of the area irrigated by canals was also served by other irrigation sources. Such conjunctive use of irrigation sources is a bright spot of irrigation. With the use of hybrid seed and chemical fertiliser technology (bio-chemical technology) spreading, the demand for dependable irrigation is intensified. Conjunctive use of water sources is an answer to growing need for dependible irrigarion.

TABLE 4.9: Sources of Irrigation — 1974-75 (Sample)

(Per cent)

Size of Holding (Acres)	Sources of Irrigation				Total
	Canal	Tanks	Wells	More than one source	
1	2	3	4	5	6
0 – 5	39.2	11.3	19.8	29.7	100.0
5 – 10	54.9	7.3	16.1	21.7	100.0
10 – 15	50.4	4.9	19.9	24.8	100.0
15 – 25	51.3	0.6	14.7	33.4	100.0
25 +	63.1	13.9	4.7	18.3	100.0
All Cultivators	52.3	8.2	15.1	24.4	100.0

Source: Field Survey, 1974-75.

In the taluka, exclusive dependence on canals seems to be relatively more for large holdings. However, there was not much variation in the percentage of area served by multiple sources for different size class of holdings with the exception of large holdings (25 acres and above). For other size classes of holdings the extent of area served by multiple sources varied between 22 and 33 per cent (Table 4.9).*

* We have used as equivalents the two terms 'conjunctive use' and 'multiple sources of irrigation'. They would be equivalent under certain circumstances. If canals supply fairly adequate water, but it is supplemented by water from other sources to gain greater control over water management, the two terms ideally match. In contrast, if cultivators at the tail-end of canals are plagued by unpredictable stoppages of water and

The extent of conjunctive use of water differs in different regions. Of the total irrigated area, over 75 per cent was served by canals in Limbasi. Limbasi thus has the overwhelming dependence on canals (Table 4.10). In Limbasi area irrigated by multiple sources was 15 per cent against 32 per cent in Matar and 27 per cent in Navagam region. In Matar and Navagam regions, canals alone in conjunction with other sources would account for 63 and 75 per cent of total irrigated area as compared to 90 per cent in Limbasi. Limbasi along with Matar region has benefited a great deal by recent expansion of canal irrigated area and has became a canal irrigated region.

TABLE 4.10: Sources of Irrigation (Regions)

(Per cent)

Region	Canals	Tanks	Wells	More than one source	Total
1	2	3	4	5	6
Navagam	48.5	18.9	5.7	26.0	100.0
Matar	31.5	0.3	36.4	31.8	100.0
Limbasi	75.0	5.7	4.0	15.2	100.0
Total	52.3	8.2	15.1	24.4	100.0

Source: Field Survey of Matar Taluka, 1974-75.

Conjunctive use of water referred to above has a limited meaning. It has operative implication at the farm level. At regional level, canals help raise water supply in wells, and hence areas not directly served by canals benefit indirectly. In Matar region wells served 31 per cent of irrigated area against 6 per cent in Navagam and 4 per cent in Limbasi. Indirect benefit of canal irrigation to Matar region could be sizeable.

---

the total supply is inadequate more often than not, additional supply of water from other sources would be to make good deficiency of canal water. The distinction is subtle yet important. It depends on the degree of dependence on canal and/or other sources of water. Dependence on two sources for the first crop (like paddy) and such dependence for the second crop like wheat can be used as one of the criteria to separate conjunctive use from the joint use. Use of two sources for paddy would be typically 'joint use'. In absence of requisite data, joint use of two or more sources has been designated by us as conjunctive use of water. In this sense the term is of course, liberally interpreted by us.

Number of Waterings

It is expected that small farmers may tend to apply greater quantity of water in order to obtain larger yields from limited land. The literature on water use however suggests that there is no advantage in using excessive water; in all probability excessive use of water leads to lower yields. The U.P. Taxation Committee has referred to such results with the help of data in regard to wheat.* In Matar Taluka we find, no relation between size of holding and number of waterings for three major crops (Table 4.11).

TABLE 4.11: Number of Waterings by Size-group, Matar Taluka

Size of Holding (acres)	Paddy	Wheat	Bajara
1	2	3	4
0 – 5	4.3	5.3	6.2
5 – 10	4.2	5.1	5.9
10 – 15	4.6	4.8	6.4
15 – 25	4.2	5.2	6.3
25 +	3.8	5.4	5.0
Total	4.3	5.2	6.1

Source: Field Survey, 1974-75.

Irrigated Crops

Of the total area under irrigation, 97 per cent was under cereals. Irrigation thus helped predominantly growing food grains. Of them, paddy and wheat accounted for 64 and 28 per cent respectively or 92 per cent, together. There were hardly any observable differences among different size class of holdings except the small holdings class (0 to 5 acres), they devoted relatively more area under *bajara* – a staple food crop in the Taluka (Table 4.12).

In Matar region though bajara is a staple cereal, paddy and wheat have emerged as dominant crops. Navagam and Limbasi devoted 99 per cent or more of irrigated area to cereals, Matar region devoted 93 per cent. Both in Limbasi and Navagam regions, paddy predominates, its share in total irrigated area is 79 and 82 per cent respectively (Table 4.13).

---

* U.P. Taxation Inquiry Committee Report, Finance Development, Govt. of Uttatr Pradesh, Lucknow. 1974.

TABLE 4.12: Shares of Crops in Irrigated Area — Matar Taluka

Size Group (acres)	Rice	Wheat	Jowar	Bajara	Other Cereals	Total Cereals	Pulses	Total Food-grains	Cotton	Any Other	Total
0 – 5	53.8	32.0	0.10	10.5	1.0	97.3	0.1	97.4	0.1	2.6	100
5 – 10	63.6	29.4	0.2	5.3	0.2	98.6	0.1	98.7	0.7	0.6	100
10 – 15	64.4	25.1	-	3.9	-	93.3	0.7	94.1	4.5	1.5	100
15 – 25	71.3	21.4	-	3.7	0.3	96.8	1.4	98.2	1.8	-	100
25 +	72.0	25.3	-	1.6	-	98.9	0.5	99.4	0.6	-	100
Total	63.8	27.7	0.1	5.3	0.3	97.2	0.4	97.6	1.4	1.0	100

Source: Field Survey, 1974-75.

TABLE 4.13: Area Irrigated, by Crops — Regionwise

Region	Rice	Wheat	Jowar	Bajara	Other Cereals	Total Cereals	Pulses	Total Food-grains	Cotton	Any Other	Total
Navagam	61.9	35.7	-	2.0	-	99.6	-	99.6	0.4	-	100
Matar	51.1	27.7	0.3	12.9	0.9	92.9	0.8	93.7	3.1	3.20	100
Limbasi	79.1	17.6	-	1.8	-	98.5	0.5	99.0	0.8	0.2	100
Total	63.8	27.7	0.1	5.3	0.3	97.3	0.4	97.6	1.4	1.0	100

Source: Field Survey, 1974-75.

Conclusion

Following observations emerge from the analysis. We find rapid expansion of irrigation. As a source of irrigation, canals predominate in all regions. Regional differentials in the extent of irrigation have narrowed. In all the three regions, over half of the cultivated area receives irrigation. While large part of the area receives irrigation in terms of protection against weather inclemency, irrigation has reached nearly all holdings. A small section, one-tenth, have no access to any source of irrigation.

Irrigation, predominantly from canals, has reached all classes of farmers. The extent of nearly full protection however is more widespread among small than among large farmers, it varies inversely with the size of holding. Fuller protection obtained to much larger extent in Navagam (old irrigated area) than in the other two regions.

Irrigation in Matar is still predominantly from one source. Conjunctive use of water is however growing, it covered nearly one-fourth of the total irrigated area in 1974-75. Such conjunctive use is more widespread in Navagam and Matar regions, and among all size classes of holdings, except the largest.

Irrigation has brought about a shift in crops in favour of cereals, especially paddy and wheat. These two crops seem to have emerged as market crops besides being used for family consumption.

APPENDIX 4.1

Gross Cropped Area and Cropped Area Under Irrigation by Size of Holding and Regions, 1974-75

(Acres)

Size of Holdings	Navagam			Matar			Limbasi			Matar Taluka		
	GCA area	GIA area	%age	GCA area	GIA area		GCA area	GIA area	%age	GCA area	GIA area	%age
1	2	3	3.a	4	5	5.a	6	7	7.a	8	9	9.a
0 to 5	378.0	244.6	(64.5)	494.5	344.2	(70.0)	221.3	119.9	(54.0)	1093.8	708.7	(64.7)
5 to 10	651.8	477.8	(73.2)	538.3	373.0	(70.0)	549.5	297.8	(54.1)	1739.6	1148.6	(66.0)
10 to 15	418.7	282.6	(67.4)	357.3	233.3	(65.2)	304.7	160.0	(56.0)	1080.7	675.9	(62.5)
15 to 25	155.8	125.6	(80.6)	103.5	88.6	(85.4)	187.1	83.4	(44.4)	446.4	297.6	(66.6)
25 +	182.3	182.3	(100.0)	47.9	20.5	(42.5)	550.5	396.4	(72.0)	780.7	599.2	(76.8)
All Cultivators	1786.6	1312.9	(73.4)	1541.5	1059.6	(68.7)	1813.1	1057.5	(58.3)	5141.2	3430.0	(66.7)

Source: Field Survey, 1974-75.

5

CROPPING PATTERN AND CROP PRODUCTIVITY

THE PRIME factor of change in the economy of the taluka in the decade of 1965-66 and 1974-75 was rapid expansion of irrigation. It influenced in a major way the cropping pattern, crop yields, production levels and intensity of land use. In this chapter, we demonstrate the impact of irrigation on land use and crop yields.

Irrigation and Crop Pattern

Study of cultivating sample households in Matar taluka reveals that between 1965-66 and 1974-75, major shifts in cropping pattern occurred. Both paddy and wheat experienced sizeable increase in their shares in the total cropped area. Percentage share of paddy increased from 31 in 1965-66 to 44 in 1974-75, and that of wheat from 21 to 33 (Table 5.1). Cereal crops dominated both in 1965-66 and 1974-75 but the composition of cereal crops differed markedly for these two years. Except paddy and to some extent wheat, cereal crops like *bajara*, *jowar* and other inferior cereals grown during 1965-66 were rainfed. In 1974-75 irrigated paddy and wheat not only replaced rainfed cereals, but they also replaced cotton which was an important crop in 1965-66 covering about 11 per cent of the total cropped area.

Data for 1974-75 provide an insight into the relationship between irrigation and cropping pattern. On the whole, for cultivating households in the sample, percentage of area under paddy was 44, on irrigated area paddy accounted for 64 per cent (Chapter 4, Table 4.13). An emphasis on paddy on irrigated area is thus visible. For wheat and for all other crops, the position differs. Share of wheat in gross cropped area in 1974-75 was 33 per cent, in gross irrigated area, its share was lower, it was 28 per cent. Thus increased irrigation in Matar tended to favour extension of paddy cultivation, by substituting other crops.

TABLE 5.1: Cropping Pattern — Matar Taluka

(Per cent)

Size Group (Acres)	Period	Paddy	Wheat	Jowar	Bajara	Other Cereals	Total Cereals	Pulses	Total Food-Grains	Cotton	Others	Total Cropped Area
0-5	1965	28.3	18.3	4.3	31.5	5.8	88.2	0.9	89.1	6.8	4.1	100
	1975	36.1	36.1	0.1	18.5	1.3	92.1	2.5	94.6	3.4	2.0	100
5-10	1965	29.6	18.3	5.7	20.8	5.1	79.5	4.4	83.9	9.3	6.8	100
	1975	42.4	35.0	0.1	34.5	1.1	93.1	3.0	96.1	3.0	0.9	100
10-15	1965	31.5	20.7	7.2	18.0	5.2	82.6	1.9	84.5	10.3	5.2	100
	1975	42.5	31.8	0.6	13.5	0.8	89.2	2.8	92.0	6.9	1.1	100
15-25	1965	35.6	21.1	7.6	12.9	3.3	80.5	5.8	86.3	8.5	5.2	100
	1975	48.5	31.4	—	15.0	0.2	95.1	3.2	98.3	1.6	0.1	100
25 +	1965	30.5	24.1	8.2	12.4	1.4	76.6	2.6	79.2	15.3	5.5	100
	1975	56.6	24.1	—	12.2	—	92.9	2.3	95.2	4.8	—	100
Total	1965	30.8	20.7	6.8	18.4	4.1	80.8	3.1	83.9	10.5	5.6	100
	1975	43.8	32.6	0.2	14.9	0.9	92.4	2.8	95.2	4.0	0.8	100

Sources: Field Survey, 1965-66 and 1974-75.

Increased dominance of a single crop would increase market sales of output. Dominance of paddy in 1974-75, both in the gross cropped area and the gross irrigated area is further evident when we look at the regional data (Table 4.14). On irrigated lands, share of paddy in terms of percentages, for three regions were 62 (Navagam), 51 (Matar), and 79 (Limbasi) (Chapter 4, Table 4.14). Share of paddy in gross cropped area was 46 in Navagam, 37 in Matar and 47 Limbasi (Table 5.2).

Yields Per Acre and Irrigation

The effect of irrigation, on crop yields, can be gauged by comparing yields on dry and irrigated area. An important preliminary insight is obtained by comparing average yields per acre for individual crops at three points of time 1929-30, 1965-66 and 1974-75. We shall bear in mind limitations of comparisons on single year yield in as much as weather plays an important part in influencing yield levels even on irrigated crops. Despite the limitations, comparison is instructive.

Three important cereals showed remarkable increases in yields ranging between 82 and 187 per cent between 1929-30 and 1965-66 (Table 5.3). These are paddy, wheat and *bajara*.

The highest increase of 187 per cent was for paddy, followed by wheat, 138 per cent. Between 1965-66 and 1974-75, paddy yield showed a modest increase of 25 per cent but wheat showed an impressive increase of 87 per cent. The highest increase in the recent decade, among cereals was that of *Bajara* – 137 per cent. Among non-food crops, cotton yields experienced highest increase between 1929-30 and 1965-66, but it had the lowest increase between 1965-66 and 1974-75. Thus there was a sharp contrast in yield changes between the period of three decades prior to 1965-66 and a decade after it. Crop like paddy, in whose favour there was a shift in crop pattern after extension of irrigation, had a modest increase in per acre yield. Wheat is another crop which benefited in area by expansion of irrigation. Increase in wheat yield was sizeable during 1965-66 and 1974-75, yet in percentage terms it was much lower compared to that observed during 1929-30 and 1965-66. These results are not necessarily pure weather effects. Both wheat and paddy get much larger irrigation protection against drought. On the whole, cereals showed greater yield increases than other crops in the first three decades. In the recent decade, barring cotton, non-food crops had sizeable yield increases.

TABLE 5.2: Cropping Pattern in Matar Taluka (By Region) (1974-75)

Region	Paddy	Wheat	Jowar	Bajara	Other Cereals	Total Cereals	Pulses	Total Food-grains	Cotton	Any Other	Total
1	2	3	4	5	6	7	8	9	10	11	12
Navagam	45.7	44.5	0.2	6.6	0.1	96.9	0.1	97.0	3.0	0.0	100.0
Matar	37.3	24.9	0.4	22.4	2.3	87.3	1.6	88.9	8.2	2.9	100.0
Limbasi	47.4	27.6	0.1	16.6	0.4	92.0	6.5	98.5	1.4	0.1	100.0

Source: Field Survey, 1974-75.

TABLE 5.3: Yield Per Acre of Principal Crops

(*kgs. per acre*)

Crop	1929-30	1965-66	Percent-age increase	1965-66	1974-75	Percent-age increase
1	2	3	4	5	6	7
Paddy	267	767	187	767	958	25
Bajara	89	212	138	212	396	87
Wheat	120	219	82	219	519	137
Inferior Cereals	216	204	—	204	606	197
Pulses	160	154	—	154	185	20
Oilseeds	84	108	29	108	361	234
Tobacco	204	226	11	226	643	185
Cotton	25	177	608	177	188	5

Sources: 1. For 1929-30 and 1965-66, Vimal Shah and C.H. Shah, *Resurvey of Matar Taluka*, Vora & Co, 1974.
2. Field Survey, 1974-75.

Comparing yield increases with cropping pattern changes, we find that crops that shared modest yield increases had larger increases in their share in the total cropped area, in the recent decade, the typical ones are paddy and wheat. As against this, crops that experienced a decline in their share in the gross cropped area, experienced demonstrable yield increases in the recent past. Three decades prior to 1965-66 were not marked by such contrasting trends in yields and area. There was a general improvement in yields and these were not related to changes in area under different crops.

These contrasting performances suggest to one conclusion: while 1929-30 to 1965-66 period was of general improvement in techniques, 1965-66 to 1974-75 period was marked by direct and relatively greater impact of irrigation on area changes. The crops whose area shrunk in size, seem to have shed relatively less suitable areas, benefit in yields in their case too might be results of cropping pattern adjustment. We may have to seek explanation for production performance in the recent past in terms of impact of irrigation on cropping pattern. However, since cropping pattern changes are linked with changes in cropping intensities we examine the irrigation and cropping intensities first and subsequently we examine their impact on the changing production levels.

Irrigation and Cropping Intensity

In 1965-66 the cropping intensity on farms of sample cultivator households was 105 only. In 1974-75 it increased to 132.8. However there were wide variations in cropping intensity both across the size of holdings and the regions.

Small holdings (0 to 5 acres) had relatively higher cropping intensity – 162.9, while large holdings had the lowest cropping intensity – 107.1. The cropping intensity and the size of holdings were inversely related (Table 5.4). The variation in intensities of land use are explained when we look at the cropping pattern once again (Table 5.1). Share of paddy in the total gross cropped area increased with the increase in the size of holding and share of wheat declined with the increase in the size of holding. In the case of small holdings (0 to 5 acres) they followed paddy-wheat and *bajara*-wheat pattern over two seasons *Kharif* and *Rabi*, medium and large holdings (10 acres and above), followed such pattern only on part of holding.

The variations also have a regional dimension. When we compare the cropping intensities across the regions – Navagam, Matar and Limbasi, we find that Navagam had the highest cropping intensity 156.4 followed by 126.30 in Limbasi and 119.2 in Matar (Table 5.5). The regional differences again conform to the cropping pattern obtained for these regions in 1974-75 (Table 5.4).

TABLE 5.4: Cropping Intensity by Size of Holdings in Sample Holdings in Matar Taluka in 1974-75

Size of holding (acres)	Cropping Intensity*
1	2
0 to 5	162.9
5 to 10	140.1
10 to 15	130.8
15 to 25	111.0
25 +	107.1
All Cultivators	132.8

$$\text{*Cropping Intensity} = \frac{\text{GCA}}{\text{NCA}} \times 100$$

Where GCA = Gross cropped area

NCA = Net cropped (or net sown) area.

TABLE 5.5: Cropping Intensity on Farms of Sample Holdings in Matar Taluka in 1974-75 (By Region)

Region	Cropping Intensity
1	2
Navagam	156.4
Matar	119.2
Limbasi	126.3
Matar Taluka	132.8

In Navagam large part of area is under irrigation for a long time. Paddy is the major crop. In Navagam, of late long duration paddy is being replaced by medium duration paddy, the process is helped by introduction of HYV seeds. Reduction in duration of paddy has facilitated growing of wheat. It may be observed from Table 5.8 that paddy and wheat taken together constituted 90.2 per cent of the total gross cropped area. The cropping pattern in Navagam seems to be paddy in *Kharif*, followed by wheat in *Rabi*. In Matar the pattern seems to be paddy and *bajara* in *Kharif* (about 60 per cent of the gross cropped area) and wheat in limited area in *Rabi*, leaving large part of land fellow during *Rabi*.

In Limbasi, however, growing wheat even without irrigation is possible because of the soil conditions and hence the intensity of land use was better than Matar in 1974-75. These observations suggest that cropping intensity depends on the cropping pattern, irrigation experience and the soil condition. We now proceed to examine changes in the production levels in the next section.

Growth of Crop Production

Impressive growth in crop production was registered between 1965-66 and 1974-75, despite the fact that crops that experienced modest yield increases had dominant shares in gross cropped area. Paddy yield increased by 25 per cent only between 1965-66 and 1974-75, but had the highest share in cropped area – 44 per cent. Increased cropping intensity and increased gross cropped area contributed sizeably to the increase in production. In 1965-66 the gross cropped area was 4434 acres with a cropping intensity of 1.05 and in 1974-75 it increased to 5141.2 acres with a cropping intensity of 1.33. Although the net cropped area declined by 8.3 per cent, the gross cropped area increased by 15.9 per cent.

For estimating growth in crop production, we took area and yield figures for individual crops. We evolved value weights for four major crops grown during 1974-75. These crops were: paddy, wheat, *bajara* and cotton. Value weights were worked out both for 1965-66 and 1974-75. Applying these value weights production indices were constructed for four crops. Results are presented in Table 5.6.

TABLE 5.6: Crop Production Indices for Sample Household in Matar Taluka

Crop	Production (in thousand tons)		Value Weights		Index for 1974-75	
	1965-66	1974-75	1965-66	1974-75	Taking 1965-66 Value Weights	Taking 1974-75 Value Weights
1	2	3	4	5	6	7
Paddy	1048	2157	0.69	0.64	142	132
Wheat	201	870	0.13	0.26	56	112
Bajara	173	303	0.09	0.08	16	14
Cotton	83	39	0.09	0.02	4	1
			1.00	1.00	218	259

Production index in 1974-75 increased at the lowest to 218 with 1965-66 value weights and to 259 with 1974-75 value weights. A geometric mean of the two indices (an ideal index – Fisher's index) would be 238. Accordingly annual compound rate of growth of crop production would work out to 9.01, 11.2 and 10.1 per cent per annum with 1965-66 and 1974-75 value weights and with the geometric mean of the two indices respectively. After making an allowance for decline in share of crops not accounted for, from 19 per cent in 1965-66 to 5 per cent in 1974-75, we get net growth rate of crop production of 7.4, 9.6 and 8.5 per cent per annum.

The analysis of the growth in crop production among cultivating sample households in Matar taluka conveys firmly that expansion in canal irrigation after 1965-66 facilitated (a) wide use of irrigation, i.e., more acreage was brought under irrigation, and (b) increased area under irrigation was sown more than once. The latter was facilitated by conjunctive use of canal and well waters. The cropping pattern changed from several alternative crop rotations practised earlier such as paddy-fallow, bajara-fallow, fallow-wheat and cotton (two seasons) covering majority of the cropped area in 1965-66 to mainly two alternatives paddy-wheat and *bajara*-wheat. Although the increase in cropping intensity was substantial, the gross cropped area increased to a smaller extent, because the net sown area declined. It thus appears that area effect resulting in increased crop production was limited. The yield effect both by way of increase in the yield of individual crops and shift in area from low-yield crops (*bajara*, inferior cereals etc.) to high-yield crops (paddy-wheat) seems to have contributed largely to the increased crop production.

6

PRODUCTION RESOURCES:
Use of Motive Power

WHILE WATER is a major input other than land and human effort, motive power, implements and machinery rank next to water in importance. Modern agriculture reflects a sea-change from its traditional stage in regard to other inputs as well. Cow-dung is supplemented to a large extent by chemical fertilisers, water lifting is done by oil engines, and electric pumps, hybrid seed provides potential for high yields. Pesticides help control of diseases through use of chemicals which replace the control through changes in genetic breeds and frequent filling to control weeds.

Use of all modern inputs needs increased investment and cash for purchase. Thus the economy of the farmer is inevitably brought into the market vortex. In subsistence economy, a farmer, as a producer, tries to push efficiency level to a high pitch to achieve maximum production level with the resources at his command. In market economy, constraint shifts from resources commanded to management ability.

Use of each modern input, involves borrowing operation, borrowing in turn would gradually unloosen from its traditional ties with landed assets. The command over resources would increase to the extent one is able to increase income and savings and use savings judiciously together with discrete operation of borrowings to push income further up. This is a process. It is slow, occasionally imperceptible, if it has to rely entirely on forces operating within the existing farming system. The process can gain speed if external forces aid. Irrigation is one such factor, especially when public sector investment brings it to the door-step of the farmer. Our endeavour here is to examine the part irrigation has played in speeding up the process of modernisation of agriculture. We deal with it in two parts. In this chapter we examine modernisation in the context of motive power and mechanisation. Impact of irrigation on use of other inputs is examined in the next chapter.

Bullocks and Tractors

Traditional source of traction power for tillage, carting and water lifting has been bullocks, bullocks have been replaced by oil engines and electric motors for water lifting, with spread of canal irrigation even this context is changed. Major use of bullocks currently is for tillage and carting. There is a change even in this limited function of bullocks. Between 1966 and 1972 for the taluka as a whole number of bullocks declined from 14,786 to 12,692. This decline resulted in increase in the land bullock ratio (in terms of land per pair of bullocks) from 15.21 acres to 17.68 acres between 1966 and 1972 (Table 6.1). Thus land bullock ratio increased by about 18 per cent in 6 years. This increase took place along with the increase in gross cropped area. Crops requiring heavy tillage like irrigated paddy and wheat also came to occupy a dominant position during the same period.

TABLE 6.1: Number of Bullocks and Area Cultivated Per Pair of Bullocks in Matar Taluka

Year	No. of Bullocks	Cultivated Area per pair of Bullocks (acres)
1	2	3
1966	14,786	15.21
1972	12,764	17.88

Source: Livestock Censuses and Season and Crop Reports.

There has been a substitution between mechanised and animal motive power. The number of tractors, for instance, increased from 47 in 1966 to 137 in 1972. Thus 90 additional tractors in 6 years were brought in. The trends observed for Matar Taluka are observed for Kheda district as well (Table 6.2).

In Matar Taluka the number of bullock-carts used for transport increased from 3,843 in 1966 to 4,603 in 1972. A similar trend in carts is noticed both for Kheda district and for Gujarat as a whole. Thus by all available indications the need for motive power supplied by bullocks has increased in the taluka, the district and the state, though their supply has declined.

Bullocks meet with competition from another side. Milch animals compete for fodder with bullocks. In Matar, total number of milch animals (cows and buffaloes) increased from 19,972 (1966) to 22,020 (1972), resulting into a decline in cultivated area per milch animal from 5.63 acres (1966) to 5.18 (1972) (Table 6.3).

In other words for given land the number of bullocks declined and the number of milch animals increased. These opposite trends in bullocks and milch animals suggest substitution between two products (crops and milk) via inputs (fodder and fodder converted into motive power). The process of such substitution seems to be pervasive, obtaining at levels of Kheda district and Gujarat State (Table 6.2).

TABLE 6.2: Number of Bullocks and Tractors in Matar Taluka, Kheda District and Gujarat State

Area	Bullocks		Tractors	
	1966	1972	1966	1972
1	2	3	4	5
Matar Taluka	14,786	12,764	47	137
Kheda District	1,92,585	1,72,636	565	2,174
Gujarat State	30,90,897	30,58,659	3,248	10,653

Source: Livestock Census, 1966 and 1972.

TABLE 6.3: Number of Milch Animals and Cultivated Area Per Milch Animal in Matar Taluka

Year	Number of Milch animals					Cultivated Area per Milch Animal (acres)
	Cows		Buffaloes (Breedable Females)		Total	
1	2		3		4	5
1966	5,550	+	14,422	=	19,972	5.63
1972	5,013	+	17,007	=	22,020	5.18

Source: Livestock Censuses and Season and Crop Reports.

Field data for 1965-66 and 1974-75 throw important light on how different classes of farmers and different regions of taluka were affected by the above-mentioned substitution process.

Number of bullocks per family declined, for different classes of farmers, by 5 per cent to 55 per cent. The biggest decline was observed for the largest size class (25 acres and above) (Table 6.4). Direct consequence of the reduced number of bullocks per farm would be increased area of cultivated land per pair of bullocks. This is observed in large measure for large farmers (25 acres and more). All other groups, except for 5 to 10 acres size of holding, also showed increased area of cultivated land per pair of bullocks. It is known widely that a large number of small farmers do not own bullocks, and they are required to pay high charges for hiring bullocks during peak seasons. Among their priorities regarding claims on their investment capacity, owned bullock would have ranked high, but for the availability of the tractor services. Very few of them can afford to own tractors. As we shall see, those who do not have bullocks take increasingly to tractor hiring for tillage operations. Small farmers would have gone much farther regarding investment in bullocks within the limited investment resources but for improved opportunity to increase the number of milch animals, and milch animals complete directly with bullocks for fodder.

TABLE 6.4: Number of Bullocks Per Farm and Cultivated Area Per Pair of Bullocks Matar Taluka

Size of Holding (acres)	Number of bullocks per cultivator household			Cultivated area per pair of bullocks (acres)		
	1965-66	1974-75	% Change	1965-66	1974-75	% Change
1	2	3	4	5	6	7
0 – 5	0.92	0.72	- 21.7	6.2	6.9	+ 11.3
5 – 10	1.40	1.33	- 5.0	10.7	10.7	0.0
10 – 15	1.90	1.38	- 27.4	12.9	15.7	+ 21.7
15 – 25	2.00	1.41	- 29.5	18.3	19.6	+7.1
25 +	3.50	1.87	- 46.6	11.3	32.4	+186.7
All Cultivators	1.45	1.08	- 25.5	10.9	12.4	+ 13.8

Sources: Field Survey, 1965-66 and 1974-75.

The data suggest sharp increase in milch cattle, and hence a steep decline in cultivated area per milch cattle is observed for different size classes of cultivators except for cultivator households with large holdings (25 acres and above). The decline ranged between 33 per cent for 10 to 15 acres size class and 64 per cent for 15 to 25 acres size class (Table 6.5). Cultivator households with large holdings experienced only a marginal increase in number of milch animals per household, whereas total cultivated area increased for this class in 1974-75. Hence the cultivable area per milch animal also increased.

The link between increased irrigation and milch animals and tractors that we can observe is that the expansion in irrigation has accelerated the two-way process of bullock substitution in the region. Small farmers have participated in the process by strengthening supply of motive power and milch animals. Large farmers have increased mechanised tractor power and underplayed expansion of dairy animals.

Dominance of tractors is brought out sharply when we examine use of bullocks and tractors. For making tractor time comparable with bullock time used in farm we have converted the tractor time into bullock-time at the rate of 1 tractor day = 4 bullock – pair days, i.e., eight bullock days.

TABLE 6.5: Area Per Milch Cattle and Number of Milch Cattle Per Farm Family

Size of holding (acres)	Year	Cultivated area per Milch animal (acres)	Number of Milch Animals Per cultivator Household
1	2	3	4
0 – 5	1965-66	5.7	0.51
	1974-75	2.3	1.09
	Change (Per cent)	- 59.6	+ 113.7
5 – 10	1965-66	8.0	0.94
	1974-75	5.4	1.31
	Change (Per cent)	- 32.5	+ 39.4
10 – 15	1965-66	12.8	0.96
	1974-75	8.4	1.29
	Change (Per cent)	- 34.4	+ 34.4
15 – 25	1965-66	21.9	0.84
	1974-75	7.9	1.75
	Change (Per cent)	- 63.9	+ 108.3
25 +	1965-66	11.6	1.70
	1974-75	17.4	1.75
	Change (Per cent)	+ 50.0	+ 2.9
All Cultivators	1965-66	9.9	0.80
	1974-75	5.4	1.26
	Change (Per cent)	- 54.5	+ 67.5

Sources: Field Survey, 1965-66 and 1974-75.

On the whole, the total tractor- and bullock-use on a farm amounted to 14.56 bullock days. Over 70 per cent of it was from the use of tractors. Predominance of tractors is observed in all the three regions, share of tractors in total bullock days ranging from 77 per cent and 75 per cent in Matar and Navagam regions to 66.2 per cent in Limbasi region.

With the exception of small farmers in Matar region predominance of tractor use as motive power is observed for all size classes of cultivators.

TABLE 6.6: Draught Power used in Matar Taluka, 1974-75

Size of holding (acres)	Bullock days used per acre	Tractor hours per acres in terms of bullock-days	Total bullock days used per acre (2 + 3)
1	2	3	4
0 to 5	6.89	7.44	14.33
	(48.1)	(51.9)	(100.0)
5 to 10	4.52	7.44	11.96
	(37.8)	(62.2)	(100.0)
10 to 15	4.07	12.56	16.63
	(24.5)	(75.5)	(100.0)
15 to 25	2.77	8.32	11.09
	(25.0)	(75.0)	(100.0)
25 and above	2.00	16.72	18.72
	(10.7)	(89.3)	(100.0)
All Cultivators	4.32	10.24	14.56
	(29.7)	(70.3)	(100.0)

Source: Field survey, 1974-75.

Small farmers (0-5 acres) had the lowest dependence on tractors 52 per cent of the total bullock equivalent days. Bullocks rapidly gave way to tractors as the size of the holding increased. The large farmers (25 acres and above) as a group used tractors to supply a little less than 90 per cent of the total traction power. This is a newly emerging phenomenon and seems to be linked directly with irrigation. Extended use of land for growing a second crop necessitates land preparation work to be done rapidly within a limited period which can only be achieved with mechanised cultivation. Limited use of bullocks and dominance of tractor use suggest that tractor has nearly fully taken over land preparation operations. It may be noted that almost all classes of farmers and all the three regions have witnessed this phenomenon. It is a major development towards mechanisation of agriculture.

Market for Tractor Services

Sample cultivator households in Matar Taluka reported both hiring-in and hiring-out of tractor services. The number of tractors owned by sample cultivator households increased from 15 in 1965-66 to 22 in 1974-75 or by about 47 per cent. It should be mentioned that this increase was much lower compared to the increase in number of tractors in the taluka as a whole within less than a decade. In the taluka as a whole number of tractors increased from 47 in 1966 to 137 in 1972 or by about 192 per cent.

From among the sample households 2.8 per cent owned tractors in 1965-66 and in 1974-75 3.8 per cent of total cultivator households owned tractors. The tractor owning households not only used them on their own farm operations, but they also hired-out tractor services. We shall now review both the aspects of tractor use, i.e., hiring-in and hiring-out.

Irrespective of the size of holding, tractor services were hired-in by all cultivators. During 1974-75 a total of 2,541 tractor hours were hired-in by all the sample cultivators. For small farmers (0 to 5 acres) hours of hired-in tractor services were the largest. However when we look at the hiring-in phenomenon per cultivator household, we find that the average use of tractor by way of hiring-in was positively related with the size of holding, i.e., higher the size of holding higher was the tractor use per cultivator household (Table 6.7).

It may be observed that although tractor services were utilised by all size class of cultivator households, large farmers, though owning tractors, had tendency to use more tractor services. Behaviour of small landholders suggest that for basic operations like preparatory tillage the tractor services seem to have been preferred over use of bullocks. Since the owning of bullocks also was not very common with this group even in 1965-66 (Table 6.4), in the changed circumstances they seem to have preferred hiring-in of tractor services instead of bullock services for tilling of land. Cultivator households owning 10 acres or more seem to have hired-in tractor services for farm operations other than preparatory tillage also.

TABLE 6.7: Hiring-in of Tractor Services in Matar Taluka (1974-75)

Size of holding (acres)	Total Tractor hours hired-in	Average Tractor hours hired-in per cultivator household*
1	2	3
0 to 5	824	3.0
5 to 10	709	4.0
10 to 15	491	6.5
15 to 25	204	7.0
25 and above	313	13.0
All Cultivators	2,541	4.6

* Note: Household in each size class excludes those who owned their own tractors.
Source: Field Survey, 1974-75.

A small group of 22 cultivators who owned tractors suggest a pattern in hiring-out of tractor services. The data are presented in Table 6.8.

TABLE 6.8: Hiring-out of Tractor Services in Matar Taluka (1974-75)

Size of holding (acres)	Number of tractors owned	Total hours hired-out
1	2	3
0 to 5	1	180
5 to 10	6	555
10 to 15	5	650
15 to 25	2	602
25 and above	8	1,056
All Cultivators	22	3,043

Source: Field Survey, 1974-75.

What emerges from the analysis is that the use of tractor is widespread and even small farmers hire-in and hire-out tractor services. In other words tractor services market in agriculture had widely developed in Matar taluka by 1974-75. We have not collected data but information has been gathered regarding forms of tractor hiring from knowledgable persons. Besides, payment on the basis of hours of use, payment in terms of share in produce obtains. Thus land tenancy is being replaced by renting of services like tractors the terms of renting being reminiscent of land tenure in a highly irrigated area. Extended use of irrigation water augments land supply. Leading to relative scarcity of complementary inputs like bullocks, which in turn are being substituted by tractors, tenancy arrangements emerge regarding a scarce production factor which is lumpy and durable.

Mechanisation and Modernisation

Modernisation of cultivation can be discussed in terms of stages of development. Initially traditional wooden plough gave way to iron plough, traditional wooden cart is being replaced by a cart with wheels with pneumatic tyres, winnowing fans came in and the crude method of waving a long piece of cloth to winnow grain gave way. Use of pesticides brought in mechanical sprayers and dusters. As tractor use spread, bullocks and ploughs gave way to tractors and cultivators (machines), use of trailors reduced the use of carts. Other attachment to tractors replaced with traditional implements like seed drill, harrow etc. In this process, change occurs in stages. We observe these stages of modernisation of cultivation and transport in Matar. Between 1965-66 and 1974-75 the process of change gathered momentum. Use of water-lifting modern devices like oil engines and electric motors is a part of the process of modernisation.

We do not have data regarding tractor accessories, attachments etc. for 1965-66. Data for 1974-75 are available. The recent data suggest that tractor accessories are owned by farmers in all size-classes. Typically, for the 22 tractors reported by sample families, 25 cultivators, 18 trailors, 7 mechanical ploughs, 13 levellers are reported (Table 6.9). Among small farmers, 1 tractor, 6 cultivators, 3 mechanical ploughs, 4 trailors and 2 levellers are reported. These numbers are not big. They only indicate that the process of mechanisation has made a head way and it has percolated to lower size classes of farmers. Importantly, ownership of accessories is not conditional on ownership of tractors, the small farmers can benefit by tractor hiring and accessory-ownership.

Modernisation in water lifting is denoted by gradual replacement of oil engines by electric motors. For the sample, total electric motors increased from 4 in 1965-66 to 18 in 1974-75 and oil engines declined in number from 51 to 39. Assuming the two did the same job with the same capacity, the total number of water-lifting devices increased

TABLE 6.9: Modern Agricultural Implements, 1965-66 and 1974-75

Size of Holding (acres)	Year	Number of house-holds	Average size of cul-tivated area in acres	Tractors	Cultiva-tors	Ploughs	Disc harrows	Levellers	Trailers	Oil engines	Electric Motors
1	2	3	4	5	6	7	8	9	10	11	12
0—5	1965	225	2.84	0	N.A.	N.A.	N.A.	N.A.	N.A.	5	1
	1975	272	2.47	1	6	3	0	2	4	8	7
5—10	1965	154	7.50	9	N.A.	N.A.	N.A.	N.A.	N.A.	13	1
	1975	177	7.02	6	6	1	1	5	4	8	3
10—15	1965	79	12.29	4	N.A.	N.A.	N.A.	N.A.	N.A.	14	1
	1975	76	10.87	5	4	0	0	3	4	15	2
15—25	1965	43	18.31	0	N.A.	N.A.	N.A.	N.A.	N.A.	9	1
	1975	24	13.87	2	0	0	0	0	1	2	3
25 +	1965	34	19.71	2	N.A.	N.A.	N.A.	N.A.	N.A.	10	0
	1975	29	30.37	8	9	3	0	3	5	6	3
All Cultivators	1965	535	7.89	15	N.A.	N.A.	N.A.	N.A.	N.A.	51	4
	1975	578	6.70	22	25	7	1	13	18	39	18

N.A. = Not Available
Source: Field survey, 1965-66 and 1974-75.

TABLE 6.10: Modern Agricultural Implements in 1974-75 (By Region)

(Number)

Region	Number of households	Average size of cultivated land in acres	Tractors	Cultivators	Ploughs	Disc harrows	Levellers	Trailers	Oil engines	Electric Motors
1	2	3	4	5	6	7	8	9	10	11
Navagam	169	6.76	10	12	3	1	4	5	19	4
Matar	252	5.13	5	5	1	-	4	6	10	9
Limbasi	157	9.14	7	8	3	-	5	7	10	5
Matar Taluka	578	6.70	22	25	7	1	13	18	39	18

Source: Field Survey, 1974-75.

marginally from 55 in 1965-66 to 57 in 1974-75. Compared to the growth of water-lifting devices, substitution between the two types is much larger, oil engines still continue to dominate in number but their domination is reduced. More than the overall situation, record of small farmers is interesting. The total number of modern water-lifting devices increased from 6 to 15, a big jump and the increase was registered by both oil engines and electric motors, steeper by the latter. Steep increase in water-lifting devices among small farmers is of special importance, the increase has been despite the rapid expansion of canal irrigation.

Mechanisation and modernisation have taken two different routes. For small farmers both traditional and modern implements owned have increased (Table 6.11). For big farmers the total number of implements owned declined, the decline is confined to traditional implements. Part of the decline in traditional implements is made up by an increase in the number of modern implements. A major change is denoted by mechanisation or tractorisation, which has replaced use of traditional implements in general. Typical illustrations are provided by plough and carts. In the case of small farmers with holdings below 5 acres traditional wooden ploughs increased from 130 to 148, iron ploughs counted separately in 1974-75 numbered 40. In the case of large farmers with holdings above 25 acres, wooden ploughs declined from 88 to 32, and iron ploughs were only 11 in 1974-75.

Similar pattern is observed with regard to carts. Traditional carts with wooden wheels increased from 54 to 81, those with pneumatic tyres from 1 to 31 – a steep increase indeed – in the case of small farmers. For large farmers both declined, the number of wooden carts from 43 to 14, those with pneumatic tyres from 8 to 6. We observe a steep increase in the number of traditional implements owned by small farmers (holdings up to 5 acres). This is evident typically in the case of seed drills, their number increased from 67 to 198. Some of the modern implements owned by small farmers include winnowing fans, chaff-cutters, sprayers etc.

For harrows no data were collected in 1965-66. In 1974-75 there were 157 harrows with 272 small farmers.

With regard to farm implements we find farmers up to 10 acres have improved their ownership of modern implements. Farmers with above 10 acres size of holding have emphasised modernisation along with mechanisation, in their case the usual implements, traditional or modern-type, declined in number.

TABLE 6.11: Traditional Agricultural Implements in 1965-66 and 1974-75 (By Size of Holding)

Size of holding (Acres)	Year	Number of households	Average size of cultivated land in acres	Ploughs		Seed Drills	Winnowing Fans	Chaff Cutters	Sprayers	Carts with Wooden Wheels	Carts With Pneumatic Tyres	Harrows	Other minor implements
				Iron	Wooden								
1	2	3	4	5	6	7	8	9	10	11	12	13	14
0 – 5	1965	225	2.84	NA	130	67	NA	NA	NA	54	1	NA	NA
	1975	272	2.47	40	148	198	3	2	9	81	31	157	2147
5 – 10	1965	154	7.50	NA	154	92	NA	NA	NA	77	5	NA	NA
	1975	177	7.02	20	139	65	4	1	5	38	15	141	930
10 – 15	1965	79	12.29	NA	118	59	NA	NA	NA	66	10	NA	NA
	1975	76	10.87	12	57	41	0	0	1	31	11	56	472
15 – 25	1965	43	18.31	NA	77	30	NA	NA	NA	39	7	NA	NA
	1975	24	16.76	9	28	20	0	0	1	12	5	25	167
25 +	1965	34	19.71	NA	88	51	NA	NA	NA	43	8	NA	NA
	1975	29	25.14	11	32	26	0	0	3	14	6	31	298
All Cultivators	1965	535	7.89	NA	567	299	NA	NA	NA	279	31	NA	NA
	1975	578	6.70	92	404	350	7	3	18	176	68	410	4014

NA = Not Available

Source: Field survey, 1965-66 and 1974-75.

Value of Implements

Small size of holding leads to two disadvantages in regard to equipment ownership. Farm operations are time specific, missing specific time for operation could lead to considerable production losses. Hence, eagerness to acquire all possible major equipments by small farmers is understandable. But this very activity leads to excessive investment.

In 1965-66 value of agricultural implements per acre in the case of small farmers (0 to 5 acres) was 80 per cent larger than that for large farmers (25 acres and above). Investment per acre of land among the middle size group of cultivators was larger than the investment made by large farmers, but it was smaller than the investment made by small farmers. In 1974-75 though the position of excessive investment in the case of small farmers continued, the differences narrowed (Table 6.12). Such narrowing is observed despite the widening base of equipment holding of small farmers, and reduction in number of major implements in the case of big farmers mainly due to heavy investment in machines by the latter. This particular point is brought out by taking value of implements and machines alone.

TABLE 6.12 Value of Agricultural Implements and Machinery Owned Per Acre of Operated Holding

(Rupees per acre in Current Prices)

Size of Holding (acres)	Value of agricultural implements, animals and wells in		Value of agricultural implements only in		Value of modern implements* in
	1965-66	1974-75	1965-66	1974-75	1974-75
1	2	3	4	5	6
0 – 5	327	1482	98	541	279
5 – 10	224	936	94	393	268
10 – 15	321	1112	184	534	444
15 – 25	269	843	109	413	324
25 +	182	937	95	649	573
All Cultivators	263	1057	118	501	373

* Modern implements include oil engines, electric motors and tractors only.

Sources: Field Survey, 1965-66 and 1974-75.

In 1965-66 per acre investment on implements alone for small and big farmers matched except for the middle size farmers, especially with 10 to 15 acres of holding, who had larger per acre investment. With the exception of this group we find, for 1965-66, per acre investment in implements was invariant with the size of holding. But this trend conceals the underlying phenomenon of inadequately equipped small farmers' group and well equipped group of large and upper middle size farmers.

In 1974-75, with widening base of implement ownership for small farmers and varying degree of mechanisation related positively with the size of holding, we find per acre investment for large farmers was the highest. For small farmers per acre investment was higher than two groups 5 to 10 and 10 to 15 acres but lower than 15 to 25 acres group. The trend in per acre investment for size group is not well-defined.

The differences in the pattern in per acre investment for size groups of land holdings with and without the pair of bullocks and wells could be explained in terms of relative emphasis on bullock ownership among small farmers. Small farmers would really miss important timing of tilling operations if they depended entirely on hired-bullocks. Some of them may have resorted to traditional exchange of bullocks and thus taking work from one bullock in place of a pair. In 1974-75 the position changed. With irrigation, mechanisation came in. Tractor services could be hired-in by small farmers. Hiring-in-of tractor services reduced the need of owning bullocks for small farmers.

Higher value of implements owned per acre of operational holding by small farmers has to be understood in the context of (a) value of implements (which are owned by all irrespective of size of holding) spreading over limited land and (b) lumpy investments in wells and water-lifting devices. We have already seen that small farmers have acquired more diesel engines and electric motors in 1974-75. Further, they also added relatively more to their milch animal stock. They score low only when modern implements like oil engines, electric motors and tractors are taken into account. Tractors especially are bulky investment which small farmers can generally ill-afford. When we consider the value of modern implements only, the position of small farmers is one rank above the lowest. For farmers with 10 acres and above value of modern implements was larger.

It should be reiterated that there is no set-pattern of the distribution of value of implements over size of holdings. The process of modernisation extends much beyond mechanisation or ownership of modern equipments. With irrigation, the use of improved inputs like fertilisers, pesticides, and high-yielding varieties of seeds increases. These inputs in fact embody a new technology. Participation of different classes in this new phenomenon is discussed in the following chapter. Improved equipments, and improved inputs together constitute modernisation of farm production processes.

7

PRODUCTION RESOURCES : Embodied Technology

Irrigation facilitated use of improved inputs like high-yielding varieties (HYV) of seeds and chemical fertilisers. Use of pesticides spreads mainly for realising full potential of new seeds and fertilisers by reducing incidence of pests and diseases.

Irrigation is a land augmenting inputs, it enables farmer to increase production of crops from a given area of land. Fertilisers and improved seeds are similarly land augmenting inputs. New varieties of seeds help in two ways. Firstly, they enable adoption of high dose, of fertilisers. Secondly, they facilitate growing of second crop by reducing duration of first crop. New seeds, fertilisers, pesticides and irrigation together embody a new technology which enables to reduce the constraints of limited supply of land.

Potential for increasing area under cultivation is non-existant in Matar taluka. In the taluka, land market operations are severally curtailed owing to abolition of tenancy and restrictions on sale or other commercial transfers of land. Hence the supply of land for an individual cultivator is also invariant to a large extent. Marginal changes can be introduced through land improvement, reclamation and development.* Use of fertilisers, HYV seeds and pesticides which are mainly purchased from the market tend to augment the supply of land.

The new technology although enables to overcome a major constraint, namely limited supply of land, it increases market dependence of cultivators for agricultural production. Market dependence requires farmers to shift emphasis from command over production resources like land to management of production resources

* These changes may increase the net sown area and cropping intensity and hence gross cropped area.

to increase incomes. In this chapter we analyse the use of land augmenting production resources and their implication for the process of marketisation.

Seeds

Use of improved seeds is known to farmers for a long time. In Matar taluka, by 1956-66 about three-fifths of the area under paddy and about two-fifths of the area under wheat was covered by improved varieties. These improved varieties embodied in them better taste or pest resistance, but not the high-yielding character. *Tinpankhali* paddy, the choicest variety, and *Jirasar,* one of the finer varieties fetched higher price than that for coarse paddy. *Niphad* variety of wheat grown in 1965-66 was a rust resistance variety.

Year 1967-68 marked the beginning of 'Green Revolution' when hybrids (where available) or high-yielding varieties were introduced for the first time in India. IR-8 (paddy) and Mexican variety of wheat were two improved strains and they were high-yielding. The role played by improved seeds in 1965-66 and by HYV seeds in 1974-75 thus is not comparable, emphasis shifted from quality to quantity of production during the intervening period. The producers responded to economic environment and technological opportunities obtaining at two points of time. The progress over the decade has to be viewed from this perspective.

Paddy is a high water-intensity crop. Improved variety seeds used in 1965-66 required relatively more water and longer care as they were long duration crops (135 days). HYV paddy used in 1974-75 required more of assured water supply. Shift from improved variety of paddy to the HYV was significant between 1967 and 1974-65. In 1974-75 area under HYV paddy among cultivating sample households was 44 per cent of the total area under paddy. Grater progress was achieved regarding wheat. Nearly one-half of the area under wheat was covered by HYV in 1974-75 (Tables 7.1 and 7.2). The spread may have been faster because wheat is a low water consuming *Rabi* crop. Two more crops that benefited by HYV are *bajra* and cotton. *Bajra* is a staple cereal. Tastes rather than yields may dominate the choice in the case of *bajra* and yet 54 per cent of the area under *bajra* was under HYV. Cotton, a commercial crop, had 39 per cent of the area under hybrid varieties.

A useful comparison can be made between the extent of adoption of improved seeds by farmers in different size-classes. The behaviour shows some distinct features. In 1965-66, in the case of paddy, size of holding indicated an advantage. Those with larger holdings adopted to a greater use of improved varieties (Table 7.1).

TABLE 7.1: Use of Improved Varieties of Seeds (1965-55) and HYV Seeds (1974-75) in Matar Taluka

(Percentage of Farmers Using Different Varieties to Total Farmers Growing the Crop)

Size of holding (in acres)	Crops in 1965-66			Crops in 1974-75				
	Paddy*			Paddy	Wheat	Bajra	Cotton	All Crops
	Pan-khali	Jirasar	Wheat Niphad					
1	2	3	4	5	6	7	8	9
0 to 5	2.4	46.0	41.8	48.5	52.0	56.5	50.0	51.7
5 to 10	13.4	47.1	44.4	40.0	50.7	52.8	31.6	46.5
10 to 15	3.1	57.8	42.3	45.5	48.5	51.2	36.9	47.0
15 to 25	2.6	84.5	38.2	40.9	34.8	31.3	50.0	36.5
25 +	3.20	87.2	15.6	33.3	40.0	75.0	14.3	41.7
All Cultivators	6.1	55.7	39.5	44.1	49.7	53.9	39.0	48.1

* These varieties are improved varieties in terms of quality — not high-yielding.

Sources: Field Survey, 1965-66 and 1974-75.

In the case of wheat, improved variety was grown largely by small and medium size class of cultivating households. The 1965-66 resurvey report states: "It is observed that improved varieties are grown as an irrigated crop. But the local varieties are largely grown as a dry crop. Most of the big farmers who grow wheat are in a region where irrigation is less prevalent."*

In 1974-75 management of resources such as irrigation, fertilisers and labour influenced the extent of adoption of HYV. Contrary to expectation, bigger size of holdings as such did not seem to endow special advantage. With irrigation common to all, we find that by and large for paddy, wheat, *bajra* and cotton small size holdings had relatively greater area under HYV. Only exception was 25+ acre size

* Vimal Shah and C.H. Shah, *Resurvey of Matar Taluka*, Vora & Co. Publishers Pvt. Ltd, 1974, Chapter 8, p.113.

class which had higher percentage of area under HYV *bajra*. With expansion of irrigation, paddy area expanded largely at the cost of *bajra* and to an extent cotton. HYV accounted for nearly one-half of the cropped area for all cultivators taken together. No significant relationship however can be observed regarding the area covered by HYV and the size of holding (Table 7.2).

TABLE 7.2: Area Under HYVs Under Each Crop by Size Class of Land Holding, 1974-75

(Percentage to total area under crop)

Size of holding (acres)	Paddy	Wheat	Bajara	Cotton	P.C. Total
0-5	46.9	45.7	72.0	62.2	51.9
5-10	49.7	48.1	59.1	34.0	50.2
10-15	52.1	42.6	69.8	49.5	51.3
15-25	42.5	43.6	48.9	75.1	44.4
25 +	39.2	75.3	76.2	18.7	51.7
All Cultivators	47.0	49.1	65.8	43.3	50.5

Source: Field Survey, 1974-75.

Fertilisers

Use of organic manures and chemical fertilizers expanded side-by-side. Data for 1965-66 are available in terms of expenditure. For 1974-75, information is available in terms of use of nutrients. The important chemical fertiliser in use during 1965-66 was ammonium sulphate. Using ammonium sulphate price, if we work out unit-cost of nitrogen, and convert subsequently rupee expenditure into nitrogen units, we get a crude estimate of nitrogen input. Nitrogen equivalent thus calculated worked out to 6.7 kg per acre. The quantity refers to purchased nitrogen. It includes both organic and chemical fertilisers. This procedure involves a bold assumption of uniform unit cost of nitrogen for all types of manures. If we accept this limitation, we get a base year quantity of nitrogen input per acre of cropped area.

If we convert for 1974-75 the total manure cost – purchased and own using area prices as the measuring unit into assumed nitrogen equivalents, we get 24.4 kg per acre, of which 19.7 kg was purchased and 4.7 kg was from own supplies. The increase in the use of purchased nitrogen equivalent amounts to 194 per cent over the decade. Most of it can be attributed to expansion of irrigation and use of HYV.

Composition of Manures

Total N P K nutrients applied to cropped area amounted to 18.6 kg (N), 4.8 kg (P), and 6.8 kg (K) per acre (Table 7.3). On demonstration farms for fertiliser trials NPK doses used in 1974-75 were 27.5 kg for N, 15.9 kg for P, and 9.9 kg for K. It is interesting to note that while major source of nitrogen was chemical fertilisers, for P and K, organic manure, i.e., dung and compost, constituted a major source (Table 7.3). Potassium (K) was almost entirely supplied by organic manures.

Compared to the dose used in fertiliser trials, actual dose applied on farms in the taluka was less by 42 per cent at the lowest. Compared to manurial doses obtaining in some of the leading Asian Nations like Korea and Malayasia, extensively irrigated agriculture in Matar trailed far behind. There obtained unexplainable regional variations in manure use (Table 7.4). Navagam with long history of irrigation, did not occupy the leading position in use of manures. Matar region led, Limbasi, with recent expansion of irrigation, understandably occupied 3rd rank but was not far behind Navagam. Thus, in the matter of manure use, irrigation history did not seem to help, current relative prices of inputs and crops would have greater influence. However, Navagam had an edge over the two regions in regard to use of both P and K. Paddy is known for depriving soil of P and K to a greater extent, their replacement with larger doses is therefore recommended by the department of agriculture. Subject to nutrient content of soil in three regions, one can observe that Navagam had a relatively more balanced use of manure.

Size of Holding and Fertiliser Use

Chemical fertilisers as a source of nitrogen supply enjoyed a prime position for all size classes of holdings (Table 7.3). In fact importance of chemical fertilisers as a source of nitrogen is inversely related to the size of holding. Absolute dose of NPK per acre of cropped area declined as the size of holding increased. While chemical fertilisers were the major source of N, organic manures were the major source for P and K for all size classes. Composition of sources of manure had little effect on dose applied, which consistently declined as size of holding increased. The 25+ acre holdings proved an exception. The P and K doses applied by them were the largest.

TABLE 7.3: Use of Fertiliser Nutrients and Their Sources During 1974-75

(kg per acre)

Size Group in acres	Dung and Compost Manure			Oil Cakes			Chemical Fertilisers			Total		
	N	P	K	N	P	K	N	P	K	N	P	K
1	2	3	4	5	6	7	8	9	10	11	12	13
0-5	7.40	3.70	7.40	0.29	0.08	0.09	18.06	1.62	0.06	25.75	5.40	7.55
5-10	5.60	2.80	5.60	1.20	0.28	0.30	12.25	2.05	0.04	19.05	5.13	5.94
10-15	3.84	1.92	3.84	0.70	0.16	0.10	11.06	1.21	0.03	15.60	3.29	3.97
15-25	3.52	1.77	3.52	0.55	0.26	0.26	9.41	0.76	0.02	13.48	2.79	3.80
25 +	12.15	6.07	12.15	0.12	0.05	0.05	3.16	0.41	0.28	15.43	6.53	12.48
All Cultivators	6.58	3.29	6.58	0.66	0.17	0.18	11.33	1.38	0.08	18.57	4.84	6.84

Source: Field Survey, 1974-75.

TABLE 7.4: Use of Fertiliser Nutrients, 1974-75

(kg per acre)

Region	Fertiliser Nutrients		
	N	P	K
1	2	3	4
Navagam	16.91	6.45	8.22
Matar	25.65	4.24	6.77
Limbasi	14.76	3.87	5.79
Matar Taluka	18.57	4.84	6.84

Source: Field Survey, 1974-75.

Complementarity and Joint Use of Modern Inputs

The use of modern inputs depends on decisions of individual farmers. Irrigation, mainly from canals, is supplied by the State and is almost in the category of an infrastructure. In a subsistence type of agriculture, irrigation is primarily an aid to overcome rain deficiency and hence is used to supplement water supplied by rain. When it begins to act as a catalyst facilitating the use of modern inputs, it becomes an instrument of modernisation. In Matar, in 1974-75, irrigation seems to have acquired this role. The progress is slow but it is in a definite direction. The evidence which we have examined so far is general. A more specific evidence obtains based on the extent of joint use of modern inputs.

Table 7.5 gives information by region and size classes of holdings regarding percentage of farmers using irrigation (I), HYV-seeds (S), Fertilisers (F) and Pesticides (P), individually or in different combinations. Data throw useful light on joint use of modern inputs. Nearly 45 per cent of all farm families in the taluka used irrigation, improved seeds and fertilisers together (I+F+S). There were 9 per cent of farm families that had no access to irrigation and did not use any of the modern inputs. I+F+S combination obtained therefore for nearly half of those farm families that used any of the modern inputs including irrigation. Those using irrigation, but not using F or S, constituted only one per cent. Next important combination of input was irrigation and fertilisers (I+F), accounting for about 20 per cent of farmers. Use of pesticides (P) in 1974-75 was found to be less popular. Hence those using I+F+S+P combination were fewer – about seven per cent only.

TABLE 7.5: Use of Modern Inputs — Individual and Joint (Matar Taluka)

Size of holding (acres)	Percentage of households reporting use of																
	I	F	S	P	I+F	I+S	I+P	F+S	S+P	F+P	I+F+ +S	I+F +P	S+P +F	I+S +P	I+S +P+F	None	Total
1	2	3	4	5	6	7	8	9	10	11	12	13	14	15	16	17	18
0-5	1.1	2.6	4.8	—	16.5	3.3	—	4.4	0.4	—	43.4	5.2	2.2	—	6.3	9.9	100
5-10	1.7	—	2.8	—	20.9	3.9	—	7.3	—	—	46.9	3.4	—	—	5.7	7.3	100
10-15	—	2.6	6.6	—	18.4	—	—	4.0	—	—	48.7	4.0	—	—	10.5	5.3	100
15-25	—	—	—	—	41.7	—	—	—	—	—	41.7	4.2	—	—	12.5	—	100
25 +	—	—	—	—	24.1	—	—	—	—	—	41.4	—	—	—	6.9	27.6	100
All Cultivators	1.0	1.6	4.0	—	19.6	2.8	—	4.8	0.2	—	45.0	4.2	1.0	—	6.9	9.0	100

I — Irrigation; F — Chemical Fertilisers; S — HYV-Seeds; P — Pesticides.

Source: Field Survey, 1974-75.

Use of any of the inputs I, F, S or P singly was less prevalent as also combinations involving P+I+S. Thus, seeds together with fertilisers or fertilisers in combination with irrigation dominated the pattern of use of modern inputs. In use, these inputs have emerged as complementary.

A few of the sample cultivating households that used I, F or S alone were small farmers, but their percentage of holdings in their own size class was insignificant. I+F+S combination was nearly equally popular in all size classes of holdings. I+F combination had a relatively greater prevalence among holdings in size classes of 15+ acres. While all cultivators taken together did not show significant inter-regional variations, the pattern of inter-size class variations in the use of input combinations showed significant variations over different regions. Percentage of holdings adopting I+F+S combination tended to decline with the increase in the size of holding in Limbasi region. Navagam showed a general trend of increase in the percentage of the use of I+F+S combination with the increase in the size of holding, an exception to this was 25+ acre size class. Small farmers are not at a disadvantage where new practices are on increase with expansion of irrigation.

Market Links and Use of Modern Inputs

With expansion of the use of non-land inputs (whether land-augmenting or not), market involvement of producers increases. Cultivators begin to hire more labour on wages and buy inputs from the market. None among the cultivating sample households was found to supply all production inputs from within the household.

A useful observation emerges if we compared paid-out costs with gross revenue. Paid-out costs including land revenue accounted for 22 per cent (Tables 7.6 and 7.7) of the total value of crop output (excluding value of by-products like straw and stumps of plants). The paid-out cost and gross revenue relationship varied within a narrow range, the percentage of paid-out costs to gross revenue being 20 per cent for small farms. Twenty-three per cent for middle group and 25+ acre size group (Table 7.6).

Relationship of paid-out cost to total cost has to be inferred. If we value labour put in by family members at the going wage rate and value bullock labour, tractor services and organic manure domestically supplied at the going price, we get an estimated value of total costs. Paid-out costs amounted to 67 per cent of total estimated cost. Details are as under :

	Percentage of Paid-out cost to Total cost
Labour : Human	19.0
Bullock	1.8
Organic Manure and Chemical Fertilisers	21.0
Seeds	6.9
Tractor Services	5.9
Irrigation	7.8
Repairs and Maintenance	2.8
Pesticides	0.9
Revenue	0.8
Total	67.0

TABLE 7.6: Share of Total Paid-out Cost to Total Value of Farm Output of Selected Crops in Matar Taluka 1974-75 (By Size Groups of Holding)

(Per cent)

Size Group (in acres)	Paddy	Wheat	Bajra	Cotton	Total
1	2	3	4	5	6
0-5	18.8	21.8	23.9	17.5	20.3
5-10	23.5	21.9	26.9	14.8	23.1
10-15	21.9	25.2	25.4	16.4	22.7
15-25	20.0	27.8	37.9	9.2	22.4
25 +	27.8	13.6	31.7	7.0	22.8
All Cultivators	22.4	21.4	26.3	13.5	22.2

Source: Field Survey, 1974-75.

Market involvement varies for different items of cost. All modern inputs are purchased from the market. Hired labour accounted for 65 per cent of total labour cost. Bullocks are usually owned. Hired services of bullocks accounted for 14 per cent of total bullock services. Tractors are recently acquired by farmers, a market for their services has recently developed (Chapter 6).

Composition of Paid-out Costs

Over one-fourth of the paid-out costs consist of hired labour (Table 7.7). Tractors, chemical fertilisers, irrigation costs, seeds, oil cakes and pesticides together accounted for a little less than 60 per cent of the total paid-out costs. In other words the market involvement for the use of modern inputs amounted to over 16 per cent of the total value of crops. This is significant proportion.

TABLE 7.7: Share of Different Items of Cost to Total Paid-out Cost, 1974-75 (All Crops)

(Per cent)

Items	Size of holding (acres)					
	0-5	5-10	10-15	15-25	25 +	All Cultivators
1	2	3	4	5	6	7
Labour	27.4	28.1	27.9	32.9	29.2	28.4
Seeds	8.9	11.2	10.7	13.5	9.3	10.4
Dung and compost manure	[illegible]	3.3	2.1	1.5	25.7*	6.2
Sheep folding	0.5	0.5	0.9	0.3	0.9	0.6
Green manure	—	0.0	0.1	—	0.1	0.1
Oil Cake	1.3	5.2	3.0	4.5	1.0	3.1
Chemical fertilisers	28.3	23.9	19.5	11.4	7.5	21.2
Pesticides	1.0	2.4	1.8	0.2	0.0	1.4
Irrigation	11.5	11.6	11.7	13.7	11.4	11.7
Repairs and maintenance	4.3	4.3	4.4	4.3	3.5	4.2
Land Revenue	1.3	1.5	1.3	1.5	1.0	1.3
Bullock-hiring	6.5	2.3	1.2	0.9	—	2.6
Tractor-hiring	6.3	5.9	15.4	7.3	10.4	8.8
Total	100.0	100.0	100.0	100.0	100.0	100.0
Cost in Rs.	333,864.0	428,291.0	288,801.0	103,216.0	206,514.0	1360,687.0

* The expenditure on purchased dung and compost is very high. This is because the use of dung and compost manure per acre of area under cultivation was 7 cart loads in Limbasi, 23 cart loads in Navagam and 4 in Matar for 25+ farmers. As against this the average use of dung and manure for other farmers was 4 carts loads. Limbasi and Navagam region account for more than 90 per cent of cultivated area owned by 25+ farmers. Thus the share of dung and compost expenditure to the total cost may be exceptionally high.

Source: Field Survey, 1974-75.

The inter-size class variations in cost composition are small. Even in the case of small farmers (0-5 acres) fertilisers accounted for about 28 per cent, tractors for 6 per cent, irrigation 11 per cent and seeds for 9 per cent of the paid-out cost. Labour hiring costs also accounted for 27 per cent of the paid-out costs. Matar experience suggests that when agriculture takes to modernisation, the extant of market involvement regarding inputs seems to be equal for all classes of farmers. As we shall see later, substantial part of wage labour is drawn from labour migrating from the neighbouring districts into the taluka, labour market thus expands beyond the local boundary.

Modernisation and Market Links : Summing-up

Over the decade, we find that with the expansion of irrigation, use of modern inputs and labour has increased. In this respect we have three major observations. In the first place Matar farmers were exposed to change in technology in the past, even prior to 1965-66 use of improved seed was known to them. The improved seeds however embodied in them qualitative improvement of products. The new seed introduced between 1965-66 and 1974-75 had a new production increasing technology embodied in it. Their use had spread almost to the same extent as that of quality seeds before 1965-66, but the progress was achieved in relatively much shorter period of about 7 years. Use of fertilisers increased, but importance of organic manure continued mainly as a source of P and K. Both in regard to use of improved seeds and fertilisers size of holding was neither a hindrance nor a help.

Secondly, we observed that combined use of irrigation, seeds and fertilisers was equally widespread among all size classes.

Thirdly, we observed that paid out costs were sizeable. They accounted for over one-fourth of the value of crops. Among paid-out costs, major items were labour, fertilisers, irrigation, seeds and repair and maintenance of implements. If we include expenses on hiring tractor services, modern inputs accounted for nearly three-fifths of the paid-out costs.

An important inference emerges from above observations. A sizeable share of purchased inputs (for current production) implies increased market involvement.

With increased marketisation of inputs, involvement in market processes is at two levels, for purchase of inputs and for sale of crop output. Firstly, for the farmer the exposure to market is more frequent. Secondly, nature of market involvement is drastically different. Buying an additional quintal of fertilisers or hiring tractor services, even additional hour of tractor services involves mentally going over the calculus of economics of production. Unlike purchase of inputs, selling surplus farm output in the market does not change the reference system of subsistence market drastically.

Yet another inference would be permitted. Not only most of the modern inputs increase market purchases but a large part of them come from outside the agricultural sector. They thus forge a link between agriculture and other sectors. The newly built bridge permits two way traffic of technology improvement. While improvement in land augmenting technology would increase the demand for non-farm sector products to be used as inputs, an improvement in production technology in the related non-farm sector would lead to widening resource use for expansion of agricultural production. This process, as our analysis shows, is linked with and is facilitated by expansion of irrigation. To sum up, expansion of irrigation transforms the farmer's approach to system of farm production and helps building inter-sectoral market links, the latter in turn reinforce inter-sectoral developmeı t efforts.

8

MARKET SUPPLY OF CROP OUTPUT

A RELEVANT development we have observed in crop production is narrowing of the range of crops grown in the taluka by sample households between 1965-66 and 1974-75. Cropping intensity also increased. Fewer crops, mainly foodgrains, grown during two seasons changed both output and marketing pattern.

Further, increased specialisation in crop production across the size classes of holdings shifted emphasis from production for consumption to production for market. A direct consequence of the pursuit of the objective of production for market was that all classes of farmers entered the market for crop sale. The traditional distinction of cash and food crops was thus obliterated. Market involvement is interlinked. Emphasis on production of crops for market would necessitate increased participation in input market. The sale of produce would thus cease to be the sale of only marketable surplus.

Disposal of Produce

Value of four major crops produced by sample cultivating households increased over the decade by 193 per cent when 1965-66 value weights were used or by 244 per cent when 1974-75 value weights are used. The growth did result in increased sales but percentage of produce sold to total production increased only marginally from 61 in 1965-66 to 64 in 1974-75. In 1965-66 about 10 per cent of crop output was used for feeding animals (Table 8.1). About 2.5 per cent was represented by addition to stock and 2 per cent by retention as seed. If these figures are added to portion retained for family consumption, total retention amounted to about 34 per cent of the total value of output. In 1974-75 such retention amounted to 27 per cent. Thus though the percentage of sale of produce in the market was over 60, and not much different for 1974-75 from that for 1965-66, retention from produce in 1974-75 represented a sizeably lower percentage. In 1974-75 payment in-kind showed however, a substantial increase over 1965-66, from 5 per cent to about 7 per cent. Remittances in kind and gift and charity accounted for about 2 per cent of total output. Thus payment in kind plus gifts, remittances etc. represented an increase which if added to retention for use, would raise retention in 1974-75 to 29 per cent. Thus the structure of disposal of farm produce changed, but the overall change in the proportion of the market supply was small.

Between 1965-66 and 1974-75 inflation was strident. Increasing portion of increased production was claimed by payment-in-kind as wages to labour mainly as a counterveiling factor. When total employment increased and migrant labour participation in labour supply increased, claims for wage payment also increased. As we have already seen in 1974-75 paid-out costs represented about 22 per cent of total value of crop output. In 1974-75, 64 per cent of the total value of output was sold, which would imply that 34 per cent of sales proceeds must have been used for meeting the total paid-out costs.

TABLE 8.1: Disposal of Crop Production in Matar Taluka, 1965-66 and 1974-75

(Per cent)

Details of Disposal	1965-66	1974-75
1	2	3
Sale	61.0	64.2
Personal Consumption	20.0	27.0
Payment-in-kind to labour	3.0	6.2
Payment-in-kind to artisans	2.0	0.6
Used for feeding farm animals	9.5	—
Gift and charity	—	0.9
Remittances in kind	—	1.1
Seeds	2.0	—
Stock	2.5	—
Total	100.0	100.0

Sources: Field Survey, 1965-66.
Filed Survey, 1974-75.

A specific structural change in crop disposal is not observed directly from the data. Between 1965-66 and 1974-75 crop production underwent a change – cotton crop designated as a market crop since total produce of cotton is usually sold, almost disappeared from the cropping pattern. Limbasi, a newly irrigated area, experienced this change in cropping pattern to a much greater extent. Paddy and wheat which were introduced in greater proportion are foodgrains. Generally, a portion of foodgrains is retained for consumption. Switch from cotton to paddy and to wheat (and also reduction in the area under *bajra* and total disappearance of *jowar*), probably led to reduced share of market sales to the total value of output.

Sale of Crops and Size of Holding

Both in 1965-66 and in 1974-75, we find that as the size of holdings increased the percentage of sales to the total crop output increased, with almost the same sharpness. For small farmers (0-5 acres) we find that the percentage of sales increased from 48 in 1965-66 to 56 in 1974-75. Thus small farmers sold over half of their output, which though high was much less compared to big farmers who sold nearly 75 per cent of their output in 1974-75 (Table 8.2).

TABLE 8.2: Sizewise Sales and Other Forms of Disposal of Production in 1965-66 and 1974-75 — Matar Taluka

(Per cent)

Size of holding (in acres)	Year	Sales	Household Consumption	Production expenses	Others	Total Production
1	2	3	4	5	6	7
0-5	1965-66	48.3	35.3	4.3	12.1	100.0
	1974-75	55.9	37.0	5.9	1.2	100.0
5-10	1965-66	57.5	25.3	4.5	12.7	100.0
	1974-75	62.2	28.1	8.1	1.5	100.0
10-15	1965-66	62.6	19.7	4.9	13.8	100.0
	1974-75	68.8	24.0	3.4	3.8	100.0
15-25	1965-66	61.3	17.1	5.6	16.0	100.0
	1974-75	71.4	20.5	5.6	2.5	100.0
25 +	1965-66	68.2	11.3	5.3	15.2	100.0
	1974-75	74.6	13.6	10.2	1.6	100.0
All Cultivators	1965-66	61.0	20.0	4.9	14.1	100.0
	1974-75	64.2	27.1	6.8	1.9	100.0

Sources: Field Survey, 1965-66.
Field Survey, 1974-75.

Foodgrains Sale

Table 8.3 gives details of sale of individual foodgrains for different size of land holdings. An important feature is conveyed by the data. On the whole in 1974-75, 77 per cent of paddy, 59 per cent of wheat and 43 per cent of *bajra* produced was sold. Paddy, thus, emerged as a major crop for market. Even small farmers with holdings below 5 acres, sold 73 per cent of their paddy output. Percentage of the paddy output that was sold increased as the size of holdings increased. For size class of 25+ acres, the sales amounted to 84 per cent of paddy output.

TABLE 8.3: Sale of Foodgrains Crops in 1974-75

Size of holding (acres)	Percentage sales to total output		
	Paddy	Wheat	Bajara
1	2	3	4
0-5	73.2	50.0	34.5
5-10	76.0	61.4	47.3
10-15	79.5	57.9	50.4
15-25	81.1	64.9	35.6
25 +	83.6	66.8	58.1
All Cultivators	77.5	58.6	43.2

Source: Field Survey, 1974-75.

Wheat is also a market crop though not to the same extent as paddy. About 59 per cent of total wheat output was sold. For wheat, the percentage of output sold increased from 50 for holdings below 5 acres to nearly 67 for 25+ acre size class. *Bajra* is a staple food. Small farmers sold 34 per cent of their *bajra* output, for big farmers (25+ acres) the percentage was 58. The sale of *bajra* was limited to 43 per cent since *bajra* continues to be a staple food crop for many households. However, the percentage of quantity marketed generally increased with the size of holdings.

Market Outlets

Table 8.4 gives details of shares of different market outlets (or

TABLE 8.4: Agencies for Sale of Crops (Quantity) by Size Class of Holdings 1974-75

(Per cent)

Size of holding (acres)	Agencies						
	Money-lenders	Coopera-tives	Govt. Levy	Millers/ Others	Commis-sion Agents	Traders	All
1	2	3	4	5	6	7	8
Paddy							
0 - 5	1.5	4.4	16.1	0.1	0.9	76.9	100.0
5 - 10	0.6	5.1	21.9	1.8	—	70.6	100.0
10 - 15	0.4	2.7	18.9	0.1	—	78.0	100.0
15 - 25	3.2	10.4	22.5	—	—	63.9	100.0
25 +	1.6	9.8	16.9	—	5.6	66.2	100.0
All Cultivators	1.2	5.9	19.1	0.5	1.2	72.1	100.0
Wheat							
0 - 5	2.0	10.2	14.2	—	0.2	73.5	100.0
5 - 10	0.7	3.9	12.0	—	0.9	82.5	100.0
10 - 15	2.6	12.0	13.5	—	—	72.0	100.0
15 - 25	—	24.9	11.1	—	—	64.0	100.0
25 +	—	9.8	5.1	—	—	85.2	100.0
All Cultivators	1.1	8.9	11.5	—	0.4	78.1	100.0
Bajra							
0 - 5	1.6	2.2	—	—	8.3	87.9	100.0
5 - 10	0.6	—	1.0	—	—	98.4	100.0
10 - 15	—	—	—	—	—	100.0	100.0
15 - 25	—	25.6	—	—	—	74.4	100.0
25 +	—	—	—	—	—	100.0	100.0
All Cultivators	0.7	1.7	0.4	—	—	94.8	100.0
Cotton							
0 - 5	—	—	—	—	38.9	61.4	100.0
5 - 10	—	18.6	—	—	21.4	60.0	100.0
10 - 15	—	—	—	—	—	100.0	100.0
15 - 25	—	—	80.0	—	—	20.0	100.0
25 +	—	57.8	—	3.9	—	38.3	100.0
All Cultivators	—	20.5	2.1	1.0	13.8	62.6	100.0

Source: Field Survey, 1974-75.

agencies) through which quantity of produce was marketed in 1974-75. Details are given for individual crops. For all major crops, wholesale traders emerged as an important agency for sale. Their share of purchase in the case of paddy was 72, wheat it was 78 and *bajra* it was 95 per cent. Traders would have had larger share even for wheat and paddy but for the fact that government levy claimed about 19 and 11 per cent respectively. Direct sales of paddy to millers or sale through commission agents or sale of money-lenders had insignificant shares. Cooperatives succeeded in acquiring a share in sale of grains and cotton. Their share varied from 5.9 per cent for paddy to 8.9 per cent for wheat and 20 per cent for cotton. Cooperatives were preferred largely by middle size or big land holders.

9

ANIMAL HUSBANDRY : Production and Sale of Animal Products

IN CHAPTER six we examined three-way substitution among draught animals, dairy animals and tractors. Tractors that substitute draught animals, relieve pressure on supply of feed and fodder and thus facilitate expansion of dairy animals. We would expect that substitution of cotton crops by paddy and wheat might add to the bulk of fodder available.

The expansion of marketing channels, through the establishment of dairy co-operatives in remote rural areas, provides assured outlets for immediate sale of a perishable commodity like milk, thus provides an incentive for increased production of milk.

Traditionally, milk production is an activity complementary to crop raising, drawing the supply of fodder mainly from plant material which is a by-product of crops raised. Dairy animals get a share in this by-product after meeting the fodder need for bullocks, which are a main source of motive power used in crop raising activity.

Hence the scope for increasing milk production in subsistence farming is limited. Fodder supply acts as a major constraint. However, when dairy is pursued mainly as an economic activity for earning income, increased milk production is obtained through improved breed of animals, improved reproduction capacity through artificial insemination, by strengthening feed and fodder supply, and by better veterinary care and improved marketing. Dairy enterprise, at that stage, enters the marketisation phase. The producer looks for economic use of feed and fodder ration so as to increase production at least cost. He enters the market both for sale of milk and for purchase of inputs. Animal husbandry then ceases to be an activity complementary to and dependent on crop production.

In this chapter we examine the problem of animal husbandry passing through a transition from subsidiary to self-reliant economic activity and enquire whether the process was triggered by massive expansion of irrigation or was induced by setting up of co-operatives for marketing milk. With limited data that we have an attempt is made to examine the process and the problem. We also examine how different occupations participated in the transition process.

Milch Animals : Growth

Table 9.1 gives two types of detail. Firstly, it gives number of milch animals per household for 1965-66 and 1974-75. For 1965-66 data pertain to farm households, 1974-75 data are available for all households including non-farm households. Secondly Table 9.1 gives details of area of operated land per milch animal owned by cultivator households. Two distinct developments are conveyed by Table 9.1. There was a substantial growth in number of milch animals and their number pressed heavily on limited size of operated land.

In 1974-75 we contacted the same households that were contracted in 1965-66, in addition we studied branched out 'satellite' households and a few new cultivator households. Thus the number of cultivator households surveyed increased from 535 in 1965-66 to 578 in 1974-75 or by 8 per cent. The number of milch animals, per cultivator household, increased from 0.8 in 1965-66 to 1.21 in 1974-75. The total increase in the number of milch animals amounted to 63.4 per cent. In 1974-75 we found that non-farm households owned about 0.46 milch animals or approximately one for every two households. The quinquennial livestock census data are available for 1966 and 1972. Taking cows and buffaloes together we find that the number of milch animals increased from 19,972 to 22,020 between 1966 and 1972 or by 10 per cent. If we take buffaloes only, their number in the taluka increased from 14,422 to 17,007 or by about 18 per cent, for 9 years the increase would amount to 30 per cent. Cows are maintained mainly by cowherds in large number and households in low income group who cannot afford a buffalo. For the taluka, number of cows declined from 5,550 in 1966 to 5,013 in 1972 (Chapter 2, Table 2.26). Since the number of households increased mainly due to branching out, the operated land per household declined and the increase in number of milch animals pressed further on limited land. Operated land per animal, for farm households, declined from 10.9 acres in 1965-66 to 5.5 in 1974-75 or by nearly 50 per cent.

TABLE 9.1: Milch Animals Owned Per Household in Matar Taluka in 1965-66 and 1974-75

Occupation Group	Number of milch animals per household			Cultivated area per milch animal (in acres)		
	1965	1975	Percentage Change	1965	1975	Percentage Change
1	2	3	4	5	6	7
Cultivator households						
Size of holding (in acres)						
0-5	0.51	1.03	+ 102.0	7.0	2.4	− 65.71
5-10	0.94	1.29	+ 37.2	9.5	5.4	− 43.16
10-15	0.96	1.30	+ 35.4	14.8	8.3	− 43.92
15-25	0.84	2.12	+ 152.4	24.0	7.9	− 67.08
25 +	1.70	1.38	− 18.8	25.0	18.2	− 27.20
Average: Cultivator households	0.80	1.21	+ 51.2	10.9	5.5	− 49.54
Non-farm households						
Agricultural labour	N.A.	0.37	—	—	—	—
Artisan	N.A.	0.26	—	—	—	—
Business	N.A.	0.14	—	—	—	—
Service	N.A.	0.15	—	—	—	—
Others	N.A.	1.59	—	—	—	—
Average: Non-farm households	N.A.	0.46	—	—	—	—

N.A. = Not Available

Sources: Field Survey, 1974-75.
Field Survey, 1965-66.

Changes in distribution of milch animals and land per milch animal across size classes of land holders are vital. Large farmers (25+ acres) reduced number of milch animals per household from 1.70 in 1965-66 to 1.38 in 1974-75, the decline was 19 per cent. Small farmers (0-5 acres) experienced a sizeable increase in number of dairy animals per household. The number increased from 0.51 to 1.03 or by 102 per cent. Except for 15-25 acres, we find the percentage increase in milch animals declined as the size of holdings increased (Table 9.1). We have no comparable data for farm labourers and other non-farm occupation groups.

Decline in cultivated land per milch animal was also lower for households with larger holdings, with the exception of 15-25 acre size class. In the three way substitution among plough animals, dairy animals and tractors, one would observe tractors enabling big farmers to reduce plough animals they held. Yet we find their enthusiasm for dairy enterprise to be low. Some of the small farmers may have owned bullocks also to hire-out their services. Despite this and despite limited opportunity to expand crop enterprise (land size being their limiting factor) small farmers seem to have shown greater interest in dairy enterprise.

Feed and Fodder

Fodder is largely obtained from own farms. Other supplies-free grazing on common property or fodder obtained as part of wages in kind in some cases also form part of the supply for some households. Dry fodder consists mainly of by-products of foodgrain plants. Farm families taken as a whole obtained about 63 per cent of the ration they fed to livestock from their own supplies. Labourers and others purchased about 78 per cent of dry fodder they used. Green fodder is also supplied from own sources, purchases being 3 per cent for farm families and 11 per cent for other families. Concentrates and other feeds are purchased from the market, own supplies constituted less than 5 per cent for farm families and 2 per cent for non-farm and farm labour families (Table 9.2). Since comparable data were not collected during 1965-66, we are not in position to observe changes.

Table 9.3 gives details of value of feed and fodder used for dairy animals. During 1974-75 a typical sample household used feed and fodder worth Rs. 637 per milch animal. A typical cultivator household used feed and fodder worth Rs. 621 and a typical non-farm household about Rs. 714. The expenditure included cost of concentrates, green fodder, dry fodder, and other expenses pertaining to veterinary services, artificial insemination and charges for servicing buffaloes. Though non-farm households used material worth 15 per cent more than that used by average cultivator household, shares of different rations in total value of feed, and other expenses were not much different. Dry fodder accounted for 48 per cent for cultivator households and 53 per cent for non-farm households. Share of concentrates was nearly the same; 24 per cent for cultivator and non-farm households. Other expenses constituted about 4.1 per cent for non-farm households, and 6.8 per cent for cultivator households. Similarly, green fodder accounted for 22 per cent for farm and 18 per cent for non-farm households.

Agricultural labourers as a class had higher expenditure level, Rs. 716 per milch animal (Table 9.3). Yet, labourers spent less on concentrates and other services. In their scheme of animal care they emphasised dry fodder – a relatively less expensive ration. Households in miscellaneous group entitled 'others' put emphasis on concentrates.

TABLE 9.2: Share of Purchased Inputs to Total Inputs

(Per cent)

Occupation Group	Green Fodder	Dry Fodder	Concen-trates	Others	Total
I. *Cultivator households*					
Size of holding (in acres)					
0-5	4.89	32.69	98.07	68.80	44.89
5-10	2.94	15.65	92.56	85.82	34.73
10-15	1.85	5.05	95.12	94.12	28.47
15-25	1.22	27.55	94.34	76.34	40.50
25 +	—	11.64	94.70	73.81	29.44
Average of cultivator households	3.01	20.21	95.33	81.14	37.10
II. *Non-farm households*					
Agricultural Labourers	12.97	76.16	99.79	96.24	71.11
Artisans	5.79	79.15	93.91	100.00	67.77
Business	19.25	62.19	100.00	100.00	63.38
Service	2.26	78.61	97.85	—	67.58
Others	12.17	77.13	100.00	100.00	72.38
Average of non-farm households	10.63	76.58	98.05	98.65	70.61
Average : All households	4.30	32.52	95.89	83.29	43.89

Source: Field Survey, 1974-75.

Levels of expenditure for artisans and salary earners were above even that obtaining for large farmers. For these two classes, expenditure on concentrates and green fodder was higher than that obtaining for all other occupation groups. As we shall see later both these classes did not have production records commensurate with input levels.

TABLE 9.3: Feed, Fodder and Other Costs per Animal in Matar Taluka 1974-75 (By Occupation Group)

(Rs.)

Occupation Groups	Green Fodder	Dry Fodder	Concentrates	Other Costs	Total
1	2	3	4	5	6
I. *Cultivator households*					
Size of holding (in acres)					
0-5	124.66(20.74)	290.10(48.27)	153.41(25.52)	32.85(5.47)	601.02(100.00)
5-10	121.77(21.04)	276.84(47.84)	135.29(23.38)	44.79(7.74)	578.69(100.00)
10-15	171.30(24.93)	313.65(45.65)	158.04(23.00)	44.12(6.42)	687.11(100.00)
15-25	136.74(23.29)	262.50(44.70)	151.16(25.74)	36.84(6.27)	587.24(100.00)
25 +	158.56(19.48)	413.36(50.80)	166.42(20.45)	75.43(9.27)	813.77(100.00)
Average of cultivator households	134.32(21.65)	295.65(47.64)	148.67(23.96)	41.88(6.75)	620.52(100.00)
II. *Non-farm households*					
Agricultural Labourers	106.70(14.90)	448.63(62.67)	139.69(19.51)	20.90(2.92)	715.92(100.00)
Artisans	332.31(21.34)	738.15(47.39)	455.00(29.21)	32.15(2.06)	1557.61(100.00)
Business	134.17(20.47)	335.00(51.11)	155.83(23.77)	30.50(4.65)	655.50(100.00)
Service	402.64(22.67)	758.64(42.72)	592.45(33.36)	22.09(1.24)	1775.82(100.00)
Others	98.02(19.30)	212.43(41.83)	157.39(30.99)	40.00(7.88)	507.84(100.00)
Average of Non-farm households	126.22(17.87)	379.12(53.07)	179.78(25.17)	29.20(4.09)	714.32(100.00)
Average of all households	132.87(20.85)	310.59(48.74)	154.23(24.20)	39.61(6.21)	637.30(100.00)

Note: Figures in brackets indicate percentage.
Source: Field Survey, 1974-75.

It is important to note that, non-farm households as a whole purchased about 77 per cent of dry fodder and 11 per cent of green fodder. Percentage of purchased dry fodder varied in a narrow range, from 76 to 79 excluding business class. Percentage of purchased green fodder varied from 12 to 19 per cent excluding service and artisan class. Thus, the basic feature of the livestock economy of non-farm households is one of close market links. The contrast between cultivating and non-farm households is marked. Among cultivator households highest percentage of purchased dry fodder was 33, it obtained for small farmers (0 to 5 acres). For the other size classes it varied from 5 (10 to 15 acres) to 27.55 (15 to 25 acres) per cent.

Thus livestock economy among sample households seems to be transistory to a full-fledged economic enterprise. As we have already observed, dry fodder accounted for only one-half of total input costs and concentrates, an indicator of better care – accounted for 25 per cent both for cultivator and non-farm households. Since share of concentrates to total value of inputs for small farmers was 26 per cent against 16 per cent for big farmers, it appears that small farmers attached relatively more importance to the dairying activity.

Production and Sale of Milk and Milk Products

We have data for milk and *ghee*. No data were obtained for curd (yogurt) and butter milk, which are consumed by the animal owning households. Butter milk is also shared with less advantaged households within the same village. Table 9.4 gives data for milk and *ghee* production, in terms of quantity and value both per milch animal and per household for different occupation groups. The data are available for 1974-75 only. The details of value of sales of milk and *ghee* are given in Table 9.5.

Total value of milk and *ghee* per household amounted to Rs. 888 for all households; for non-farm households it was very low, Rs. 369; for cultivator households, it was Rs. 1367. For cultivator households, gross value of crops produced per household amounted to Rs. 8,113 at 1974-75 prices. Thus, milch animal products amounted to about 17 per cent of the value of crop output. Value of milk and milk products sold by cultivators was Rs. 782, which was about 15 per cent of the value of crops sold. Contribution of livestock products to the value of the crop output varied in a wide range. Percentage of value of output of dairy products value of crop output was 22.2 for small farmers and between

16 and 18 per cent for size classes up to 25 acres. Large farmers (25+ acres) had in fact much lower sales, dairy products accounted for 6 per cent of value of crop output.

TABLE 9.4: Output and Value of Milk and Milk Products in Matar Taluka, 1974-75 (by Occupation Group)

Occupation Group	Milk Quantity (Litres) Per Milch animal	Value (Rs.) Per Milch animal	Ghee Quantity (kg) Per Milch animal	Value (Rs.) Per Milch animal	Total value of products (Rs.) Per milch animal
1	2	3	4	5	6
I. *Cultivator Households*					
Size of holding (acres)					
0-5	567	998	1.03	19	1017
5-10	615	1115	2.01	52	1167
10-15	735	1369	3.00	75	1444
15-25	618	1108	2.29	51	1159
25 +	407	742	7.00	173	915
Average of cultivator households	601	1082	2.07	49	1131
II. *Non-farm households*					
Agricultural Labourers	512	871	0.81	17	888
Artisan	421	663	3.00	90	753
Business	562	1106	—	—	1106
Service	626	1168	—	—	1168
Others	420	668	—	—	668
Average of non-farm households	477	798	0.51	12	810
Average of all households	569	1009	1.67	39	1048

Source: Field Survey, 1974-75.

Two observations can be made on the basis of the above findings. Share of value of output from milch animals to the value of crop output or crop sales is significant though not large. Secondly, for small farmers it is an important source of income.

TABLE 9.5: Disposal of Milk and Milk Products in Matar Taluka, 1974-75 (By Occupation Group)

(Per cent of Output)

Occupation Group	Milk (in Liters)				Ghee (in Kg)		
	Home Consumption	Sold to Co-op Society	Sold to others	Total	Home Consumption	Sold to Others	Total
1	2	3	4	5	6	7	8
I. *Cultivator Households*							
Size of holding (in acres)							
0-5	26.0	70.7	3.4	100.0	85.1	14.9	100.0
5-10	28.7	66.8	4.6	100.0	86.1	13.9	100.0
10-15	27.2	67.3	5.5	100.0	78.8	21.2	100.0
15-25	35.3	56.3	8.4	100.0	94.9	5.1	100.0
25 +	58.6	27.1	14.3	100.0	98.6	1.4	100.0
Total : Cultivator households	29.1	66.0	4.9	100.0	87.6	12.4	100.0
II. *Non-farm households*							
Agricultural Labourers	27.2	66.6	6.2	100.0	67.0	33.0	100.0
Artisan	54.9	38.6	6.5	100.0	53.3	46.7	100.0
Business	32.8	62.5	4.7	100.0	—	—	—
Service	36.6	63.1	0.3	100.0	—	—	—
Others	26.4	62.6	11.1	100.0	—	—	100.0
Total: Non-farm households	28.6	63.8	7.6	100.0	63.7	36.3	100.0
All households	29.0	65.5	5.5	100.0	85.7	14.3	100.0

Source: Field Survey, 1974-75.

A similar comparison will be useful for non-cultivating families. Sources of income for each group varied. It is our belief that most of the non-cultivating families, who do not have fodder supply of their own, pursue livestock economy as an enterprise. This is corroborated by market sales of milk and *ghee*. On the whole 71 per cent of milk produced was disposed of through sales. Both cultivators and non-cultivators had a comparable percentage. But sale percentage varied widely among sub-groups of cultivators and not among non-cultivating households. Except for artisans who sold 45 per cent of milk output, all other classes including wage labourers sold about two-thirds of milk output. We have already observed that non-cultivating households purchased over three-fourths of their feed and fodder requirements. For cultivating families percentage of sale of milk to milk output was 74 per cent for small farmers and 41 per cent for large farmers. Producing milk for market is thus a dominant feature of all occupation groups, more so of non-cultivating families and small farmers among cultivatory. Unlike milk, *ghee* is produced only for self-consumption. Artisans and agricultural labourers were exceptions, in their case percentage of *ghee* sold to the total *ghee* output was 47 and 33. Among cultivators, percentage varied from 1. 4 (big farmers with 25+ acre) to 21 (10-15 acre). The small farmer sold 15 per cent of ghee output (Table 9.5).

On the whole, per milch animal output value was higher for cultivating families but business and service classes attained the same level as the cultivator households. Among the latter classes middle size groups (5 to 25 acres) had relatively higher level of output per milch animal.

10

LABOUR: Employment and Wages

IRRIGATION HAD a direct impact on labour employment. Irrigation itself helps raising labour absorption. With crop pattern changing in favour of paddy in Matar Taluka, increase in labour absorption in crop production was fairly steep. Whether increased opportunity for employment results in increased wage income would depend on response of labour supply. In Matar, new opportunity attracted migrant labour from a neighbouring district. Labour market as a whole experienced a change in regard to the length of employment period during the year. Terms and conditions of work also changed.

As shown in Chapter 2, Matar Taluka experienced an increase of 30.4 per cent in total population between 1961 and 1971. The estimated growth of population between 1965-66 and 1974-75 was 23.4 per cent. There was a change in definition of workers between 1961 and 1971. The basis of defining a worker changed from income to time disposal. Using the correction factor evolved by the census, we find an increase in workers in Matar taluka of 24.8 per cent between 1961 and 1971.

Between 1965-66 and 1974-75, percentage of workers to total population increased from 37 to 41 for cultivators and from 52 to 54 among farm labour families. Further, households reporting agricultural labour as main occupation of the head were 24.4 per cent in 1965-66 and 28.3 per cent in 1974-75. The increase in agricultural labour households was 15.44 per cent. From among agricultural labour households, landless households in 1965-66 were 78.7 per cent, in 1974-75 they constituted 84.7 per cent. Increased proportion of agricultural labour households, and among them of landless households represents, swelling of ranks of those available for work on wages.

Increased supply of wage earners may also be due to reduced percentage of these households reporting other subsidiary occupations. Against 25 per cent reporting other subsidiary occupations in 1965-66, in 1974-75 only 13 per cent reported subsidiary occupations.

Some important features of agricultural labourers need to be noted. In the first place from among those reporting subsidiary occupations in 1965-66 a few (8.7%) reported land holding above five acres, in 1974-75 none reported owning land more than 5 acres. Besides, the percentage of those reporting subsidiary occupation other than farming declined on the whole. Thus most of the wage earners depended increasingly on agricultural employment.

Second important feature of agricultural labour household was that number of members per household marginally declined from 4.4 in 1965-66 to 4.3 in 1974-75, but number of workers per household remained unchanged, 2.3 per family. Even ratios of workers to total population of adult males and females and boys and girls do not seem to have changed significantly. On the supply side a major change had occurred, it pertains to immigrant labour. They accounted for around two per cent of total labour on farm in 1965-66. In 1974-75 their proportion increased to 14 per cent and they accounted for over 21 per cent of wage labour (mandays) put in on farms.

In 1974-75 local wage labour accounted for 31 per cent of total workers engaged in agriculture, 85 per cent of adult males and 62 per cent of adult females from the labour households were workers. In 1974-75, relatively larger proportion of females and males participated in work from among farm labour households and hence the overall participation among these households was higher than the participation from among cultivating households (Table 10.1).

TABLE 10.1: Labour Participation Rates in 1974-75 of Sample Families by Size Groups of Holdings

(Per cent)

Occupation	Workers' Participation 1974-75 (All Age groups)		
	Males	Females	Total
Cultivators			
0 to 5 acres	60	31	46
5 to 10 acres	54	22	39
10 acres and above	52	16	34
Total Cultivators	56	25	41
Agricultural Labour	60	48	54
Total	57	31	45

Source: Sample Survey, 1974-75.

Among agricultural labourers nearly two-thirds of males and one-half of females offered for work taking all age groups. For cultivators, on the whole, proportion of males engaged in work was only slightly below that of agricultural labour, from among females about a quarter engaged themselves in work, against nearly half among agricultural labourers. In regard to work participation patterns, for males, agricultural labourers compare with small farmers, for other farmers we find that the participation declined as the size of holding increased. In the case of female workers the participation rate was low among cultivator households, the rate further declined with the increase in the size of holding.

Labour Absorption

Judged from various angles we find that supply of labour in agriculture increased sharply, the increase was steep in the case of wage paid labour. The migrant labour also came in increased proportion. When we juxtapose demand for labour against its supply, we find that former increased equally sharply if not more. We have data regarding labour absorption in detail for 1974-75. We do not have similar data for 1965-66. With the use of data for 1974-75, an inference can be drawn, albeit crude, regarding the magnitude of the change.

Table 10.2 gives for 1974-75 mandays employed per acre for different crops.

TABLE 10.2: Mandays per Acre by Crop and Irrigation Status in 1974-75: Matar Taluka

(Mandays/acre)

Crop	Irrigation Status of holding		
	Unirrigated*	Irrigated	Per cent difference
Paddy	41.21	45.60	11.07
Wheat	15.05	29.13	93.55
Bajra	19.75	42.47	113.52

* Unirrigated holding here refer to those holdings whose 33 per cent or less area had access to irrigation during 1974-75. Since most of the holding had about 50 per cent of the area with access to irrigation, the unirrigated holding was defined as mentioned above.

Source: Field Survey, 1974-75.

A straight comparison of mandays absorbed per acre between irrigated and unirrigated crops shows an increased absorption of labour ranging from 11.1 per cent (paddy) to 93.6 per cent (wheat) and 113.5 per cent (*bajra*). Between 1965-66 and 1974-75, irrigated area increased and cropping pattern changed in favour of paddy and wheat. A comparison of labour input in unirrigated *bajra* with irrigated paddy would amount to 131 per cent increase which is substantial. As already noted in Chapter 5 area under *bajra* declined from 18.4 per cent to 14.5 per cent and that under paddy increased from 31 per cent to 44 per cent between 1965-66 and 1974-75. Area under wheat increased from 21 to 32 per cent. Wheat absorbed about 10 per cent less mandays than what *bajra* did. However combination of paddy in *Kharif* and wheat in *Rabi* would lead to a substantial increase in labour, employment increase could amount to three times on per acre of net area sown.

Total mandays per acre of gross cropped area, weighted on the basis of 1965-66 area weights worked out to 19.8. On the basis of 1974-75 area weights, they worked out to 31 mandays. The increase amounts to 57.2 per cent. Regionwise mandays absorbed per acre in 1965-66 and 1974-75 are presented in Table 10.3:

TABLE 10.3: Mandays Per Acre of Crop Production in 1965-66 and in 1974-75: Matar Taluka and Regions: Mandays/Acre

Region	Mandays per acre		Change (Per cent)
	1965-66	1974-75	
Navagam	29.45	31.00	5.3
Matar	14.40	29.71	106.3
Limbasi	13.15	31.37	138.6
Taluka	19.79	31.11	57.2

Source: Field Survey, 1974-75.

In 1965-66 use of chemical fertilisers was limited, high-yielding varieties had not been introduced, area under irrigation was less. Considering these facts we would infer that in 1965-66 level of per acre employment for *bajra* and cotton could be less than that observed in 1974-75. In 1974-75 paddy and wheat occupied nearly half and one-third of cropped area respectively. Considering these facts it could be inferred that there would be a substantial increase in labour absorption. This could be particularly observed in the case of Matar and Limbasi regions, where a shift to paddy-wheat combination took place in a big way in the recent decade.

Seasonality

A steep increase in labour absorption would not indicate directly the pressure of demand on supply unless the seasonal distribution of labour at two points of time had remained unchanged. The seasonal pattern has significantly changed, it has been influenced by the change in cropping pattern and change in irrigated area.

Total duration of traditional *bajra* is around 105 days and that of HYV paddy is 115 days. Paddy has three major operations, transplanting, weeding and harvesting, they absorbed 12, 8 and 14 mandays per acre, which were nearly 78.7 per cent of total mandays per acre reported for irrigated paddy. *Bajra* has two major operations, weeding and harvesting and they absorbed 68 per cent of total mandays per acre (Table 10.4). In other words, for nearly comparable duration total employment with greater intensity at two points of time during *Kharif* season has been replaced by larger total employment with three points of labour absorption intensity. Peak in labour employment in paddy was nearly twice as high as that in *bajra*. We should note that in 1974-75 hybrid *bajra* came in use, the duration of which was reduced to 80 days. Duration of paddy too was reduced from 141-145 days, to 110-115 days. Reduction in duration of crops made seasonality sharper while facilitating growing of two crops on the same land. At any point of intense demand for labour, labour employed in 1974-75 would be much larger than that in 1965-66. Besides, there would be less frequent periods of lack of employment. Thus the changed pattern of seasonality would also mean more sustained work for those participating in work. Part-time or marginal workers would or could withdraw. Such withdrawal would be mostly in the case of land holding cultivators. For wage earners, sustained work for longer duration stretching far out in *Rabi* season with increased overall demand, means rising opportunity for employment.

Hired Labour Employment

In a traditional economy, most of the labour is supplied by the family. Unpaid labour, which includes exchange labour would also be predominant. Matar taluka seems to have passed this stage. In 1974-75 even small farmers hired 57 per cent of total mandays used and big farmers hired about 80 per cent of total mandays used for growing paddy. For paddy we have noted three major seasonal operations, small farmers employed relatively more hired labour for transplanting and weeding but big farmers employed over 80 per cent of hired labour for all the three major operations (Table 10.5). If we consider a shift from *bajra* to paddy, demand for hired labour would indicate an increase of over 157 per cent. Increase in hired labour would amount to a little more than 97 per cent, when unirrigated wheat is substituted by irrigated wheat.

Migrant Labour

Increased demand for labour attracted migrant labour mainly from the adjoining district of the Panchmahals. While overall migrant labour represented 2 per cent of total labour absorbed in 1965-66, in 1974-75 about 14 per cent of labour put in was supplied by migrant labour. Migrant labour accounted for about 25 per cent of hired labour. Migrant labour came in mostly during monsoon. Eighty-seven per cent of the labour put in by migrant labour was for *Kharif* crops, 11 per cent was for winter crops and only 2 per cent for summer crops. Migrant labourers were hired to a large extent by large farmers. On holdings below 5 acres in size, migrant labour put in 18.3 per cent of total labour. On holdings above 10 acres they accounted for 32.9 per cent of total labour.

TABLE 10.4 Absorption of Mandays Per Acre of Unirrigated *Bajra* and Irrigated Paddy

Agricultural Operations	Mandays per acre for	
	Unirrigated Bajra	Irrigated Paddy
Prepatory tillage	2.1	2.9
Sewing, transplanting	1.7	11.6
Interculturing	0.9	1.5
Weeding	6.7	7.9
Irrigation	0.2	2.3
Manuring	1.0	1.7
Pesticides application	0.0	0.6
Harvesting and post Harvesting operations	7.8	13.9
Total	21.4	42.4

Note: The difference between the total mandays per acre by operations do not tally with the figure in Table 10.2 because the figures for mandays by operations are averages for each operation whereas the figures in Table 10.2 show the average use of labour for all operations taken together.

Source: Field Survey, 1974-75.

TABLE 10.5: Percentage of Hired Labour to Total Labour used by Size Class of Holding, Irrigation Status and Crops

(Per cent)

Size of holding (acres)	Crops				
	Paddy	Bajra		Wheat	
		Unirrigated	Irrigated	Unirrigated	Irrigated
1	2	3	4	5	6
Less than 5	57.2	45.9	53.9	49.5	54.3
5 to 10	68.1	54.2	51.9	55.6	67.1
10 and above	79.8	71.3	53.1	82.8	69.7
All	70.3	56.0	53.0	62.7	63.7

Percentage of hired labour to total labour used by size class of holding, irrigation status and crops.
Source: Filed Survey, 1974-75.

Limbasi and Matar regions experienced rapid expansion of irrigation in the recent past. Share of migrant labour in total labour differed distinctly regionally. In Limbasi, migrant labour accounted for 26.3 per cent of total mandays and 42 per cent of hired labour, while hired labour as such accounted for 63 per cent of total mandays. Share of hired labour to total labour ranged between 63 (Matar) to 67 per cent (Navagam) but share of migrant labour varied from 5 per cent (Matar) to 18 per cent (Navagam) and 26 per cent (Limbasi).

Thus Matar region employed mostly local labour and Limbasi employed migrant labour to a greater extent. On rush of migrants would lead to increased competition between migrant and local labour and where the former proved superior to some section of local labour, a situation of 'pools' of unemployed may emerge amidst rising employment opportunities.

TABLE 10.6 Percentage of Migrant Labour to Hired Labour in Matar Taluka by Size of Holding

Size of holding (in acres)	Migrant labour as per cent to total labour absorbed	Migrant labour as per cent to hired labour
Small (0 to 5 acres)	10.8	18.8
Medium (5 to 10 acres)	14.5	20.9
Large (10 acres and more)	28.4	32.9
All	14.0	25.0

Source: Field Survey, 1974-75.

TABLE 10.7 Migrant and Hired Labour (Regions) in 1974-75

(Per cent)

Region	Migrant labour as per cent to total labour absorbed	Migrant labour as per cent to hired labour	Hired labour as per cent to total labour
Navagam	18.9	28.2	67.1
Matar	4.8	7.6	63.1
Limbasi	26.3	41.7	63.2
Matar Taluka	14.0	25.0	66.0

Source: Field Survey, 1974-75.

Types of Labour

For 1974-75 we get details regarding male, female and child labour and among wage labour those employed on permanent or annual basis, those engaged on piece basis and those engaged on a daily basis. Details pertaining to leading crops and to holdings small (0-5 acres) and large (10 acre and above size group) are presented in Table 10.8.

Use of casual labour predominates for both paddy and wheat dry and irrigated: the exchange labour had the lowest share, this arrangement seems to be almost on way out. Weather uncertainty and time specificity requires an emphasis on assured labour supply within a short period, and hence preference for contract labour or piece labour in unirrigated wheat mainly among large farmers. For irrigated wheat, it seems continued and assured supply of labour is important especially for large farmers and hence they hired farm servants, who contributed nearly one-fifth of total labour.

Wages and Wage Payment

Wages are paid in cash as well as in kind. The practice of wage payment has changed over time. It is in all probability linked with prevailing inflation and employment situation. Back, in 1929-30, of the total payment of wages to attached and other labour 79 per cent was in cash, but in 1965-66, 57 per cent was paid in cash and the rest in kind. In 1974-75, the contracts of labour employment changed. Contract or piece labour and annual servant accounted for 80 per cent of total hired labour for paddy. In the case of large holdings for wheat they accounted for 70 per cent. Contract labourers were given wages in cash along with other perquisites like fuel, site to pitch tents for temporary accommodation etc. The annual servants were given perquisites in kind like meals, clothes, *bidies*, beverages and occasional medical aid. Casual labour is given a cup of tea, one meal, and *bidies*. These practices have been unchanged over the decade between 1965-66 and 1974-75. We find for casual labour, payment of wages were predominantly in cash, about 85 per cent, and 15 per cent in kind. These payments do not refer to perquisites. Proportion of wages in cash and kind differed for different crops. For a cash crop like cotton, only 2 per cent of wages were paid in kind, for wheat in-kind component amounted to 19 per cent, for paddy 15 per cent and for *bajra* 11 per cent. If the same practices as observed in 1974-75 obtained also in 1965-66, the cash component of wages could be higher for casual labour, since in 1965-66 cotton and *bajra* occupied a larger percentage of cropped area. If we combine casual labour and annual servants (attached labour) and add perquisites to wages, the cash component would be about 67 per cent. Thus we do not observe a specific trend indicating a change in wage practices. The details for 1974-75 are given in Table 10.9.

Real Wages

Since wages are paid in cash to a greater extent, with steep rise in prices, and with absence of labour organisation for collective bargaining, wages in cash may move up with a lag and hence the real wages may even tend to decline. Such a trend is most likely if labour supply rather than demand for labour press on equilibrium wage or unemployment is widespread. If wages are already rock-bottom permitting only subsistence, a decline in real wage is less likely. Protection of real wage is provided in the case of subsistence wage through dominance of wages in kind.

Data regarding wage rates are scanty. For 1965-66 for paddy transplanting and weeding, wages were Rs. 2.50 and Rs. 2.00 per day respectively. In 1974-75 wages paid in cash for paddy for males were Rs. 4.40 for transplanting and Rs. 3.60 for weeding. Thus cash wages in money increased by 76 and 80 per cent for transplanting and weeding in the case of paddy. During the period between 1965-66 and 1974-75 the paddy prices increased by 67 per cent. The real wages thus have tended to increase though increase could be marginal. To find the underlying trend in real wages we should in fact compare wages in *bajra* against wages in paddy, for the latter has replaced the former in newly irrigated areas.

TABLE 10.8: Types of Labour Absorbed in Paddy and Wheat Crops by Size Classes of Land Holding and by Irrigation Status, 1974-75

(Per cent)

Components of Labour	Paddy				Wheat			
	Small Holdings		Large Holdings		Small Holdings		Large Holdings	
	Unirri-gated	Irriga-ted	Unirri-gated	Irriga-ted	Unirri-gated	Irriga-ted	Unirri-gated	Irriga-ted
1	2	3	4	5	6	7	8	9
Permanent Labour as % to total labour used	6.06	5.43	12.57	9.13	13.43	4.24	17.67	21.59
Casual Labour as % to total labour used	31.17	46.86	48.70	62.77	30.32	46.69	28.31	42.87
Casual female to total casual labour	34.91	19.76	16.49	16.17	26.04	17.85	25.82	17.48
Contract Labour as % to total labour used	15.21	4.14	15.14	8.94	11.58	3.32	27.29	5.25
Hired Labour as % to total labour used	52.44	56.45	76.91	80.86	55.33	54.25	73.26	69.71
Total exchange labour as % to total labour	3.59	5.05	1.34	1.80	1.76	2.87	–	3.92

Source: Field Survey, 1974-75.

TABLE 10.9: Payment of Wages to Casual Labourers by Crops (1974-75)

(Per cent)

Crop	Wages paid in	
	Cash	Kind
Paddy	84.7	15.2
Wheat	81.4	18.6
Bajara	89.0	11.0
Cotton	97.9	2.1
All Crops	84.9	15.1

Source: Field Survey, 1974-75.

Paddy wages in 1965-66 would reflect largely wages prevailing in Navagam. In Limbasi, *bajra* predominated in 1965-66, between 1965-66 and 1974-75 major change in cropping pattern occurred in Matar and Limbasi regions. In Limbasi again, labour shortage attracted migrant labour. In 1974-75 paddy predominated in all regions, but in Navagam where expansion of irrigation was limited and expansion of area under paddy was minimal, relatively smaller increase in wages would be observed. Labour is mobile and therefore we do not expect inter-regional wage differentials to be marked in 1974-75, besides in 1974-75 migrant labour had a larger share in total labour. In 1965-66 such was not the case. Low immigration could be accompanied by low internal migration (within the taluka) and we would expect greater inter-regional wage differentials in 1965-66. Weeding is mostly for *bajra*, transplanting labour is specific to paddy, hence a cross comparison of wage rate for weeding in 1965-66 with transplanting wage rate in 1974-75 would be indicative of a probable change in wage rate. Such a comparison shows 120 per cent increase in money wage rates. This would imply 50 to 70 per cent increase in real wage rate mostly in newly irrigated regions.

Analysis of data for labour employment and wages suggest that, during the decade under study, overall employment level has tended to increase, the increase is more for hired labour, among them for casual labour. The share of migrant labour, coming mostly form adjoining district of the Panchmahals increased impressively they found employment in newly irrigated region of Limbasi to a relatively greater extent. A major change was in the nature of wage contract, on the whole piece rate contract and annual contract together came to dominate. The available evidence points to a probable increase in real wages despite large-scale immigration of labour. Though competition between local and migrant labour could put the local - at least some sections of it and in some areas - to a disadvantage, leading to small pools of unemployed local labour, juxtaposed to higher employment level.

11

ASSETS, LIABILITIES, SAVINGS AND INVESTMENT

IRRIGATION ACTS as a catalytic agent for growth. Not only does irrigation help increasing production, it stimulates growth by encouraging investment in productive assets. We have examined in foregoing chapters growth of production, use of inputs and change in technology. We have also examined inter-relations among them. We examine in the present chapter changes in level and structure of assets, their relation to investment and liabilities. We include in assets land as well. We postulate that along with expansion of irrigation, when production technology changes with the use of bio-chemical inputs, potential is created for further investment in physical durable capital. On the supply side, increased income makes possible to release larger resources for investment. Along with own resources, income transfer from future to present is facilitated by expanding activities of lending. All in all, investment activities build up rapidly. Not all of them are directly related to utilisation of increased irrigation facilities, the latter plays mainly the role of a catalyst. What applies to physical durable capital, applies equally to human capital formation. However, in this chapter we confine our analysis to the physical durable capital only.

Physical durable capital is a well defined concept. However in accounting for assets covered by the concept we run into a problem regarding two specific items. Among them land is one. To the producer land is as much an asset as any other physical asset, say like a bullock. Even if land is inherited it represents, a stock of value for which there is a market. Flow of services however can be conceived for 'infinite' time period as the depreciation of 'Land as an asset' is minimal, even imperceptible. Such an argument conceals a major role of land in farm production. Land, from production point of view, is a store of crop nutrient, which gets depleted by raising crops. A fine balance between nutrient deposition from air (and water) can be maintained at a low level of production activity. When production activity increases – as in the case of irrigated crops like paddy – replenishment of depleted nutrient is called for. Fertilisers help restore the depleted stock of nutrients. With increased use of fertilisers, role of land as a store of soil nutrients is diminished and along with it its relative value. But, its other role gains in prominence. Other role of land is to provide space for crops to stand on, and when number of plants bearing larger harvest, stand on it, its value of land as supplying space for crops to grow increases. Depletion of flow of space is negligible, except when land is transferred from crop production to other uses. This does happen, with expanding economic activity, use of land for non-agricultural purposes expands. There is yet another aspect. Quality of land as space needs maintenance and investment over a period of time. Erosion, creation of gullies, waterlogging and salinity reduce even value of land as space unless care is bestowed on it. In this particular sense land acquires the status of an economic asset. The role is not however as clearly defined as that of bullock. Depreciation is non-measurable and maintenance as apart from investment is indeterminate. We take a view that land may be accounted as an asset, but building of other assets may be considered separately as well.

House is yet another asset the category of which is not clearly defined, but for a different reason. A useful distinction needs to be made between physical durable capital for production and that for consumption. A set of furniture pieces, or utensils, are useful for housekeeping, its services to production are not direct and essential. Is house an aid to production? Views differ. In a nuclear village, house would have less direct relevance of its location to facilitate supervision of production activity. However it does provide storage space for implements, bullocks, seeds, inputs like fertilisers and also for the produce. It houses animals too. Quality of house is of secondary importance from storage point of view. The difference between a mud house and a *pucca* walled house, represents largely a 'taste' for consumption of 'housing' or a house as a dwelling. House has therefore an undefined combination of production and consumption asset components. We therefore consider it for analysis, along with as well as apart from, other productive assets.

Growth of Assets

Measurement of growth of assets raises important problems. Some of them are methodological, others are of measurement. Methodological issues pertain to identifying growth on the basis of investment, or cumulation of assets from year-to-year, added on to the base year stock. Alternatively, we take a difference between the inventory of the stock at two points of time. Conceptually, two methods should give identical results. Depreciation is notional but replacement is a real phenomenon. Empirically, no account is available, not even for Matar, for expenditure on replacement. The two methods used to measure growth of assets could therefore give empirically two different results, cumulation method will give an upward bias to gross investment if no account is taken of depreciation. The price variation will be an additional complication. In contrast, the inventory method will give net investment. Bias, will be undefined since distribution of replacement expenditure over time span may be well behaved or may be bunched and peaked. In a growing economy if quantitative or qualitative (vintage) improvement is rapid, in both the cases the inventory method will give a lower bound. We have preferred inventory method mainly because of limitation of available data. We, however, analyse investment additionally.

The measurement problem, even for inventory method, is complex, we intend to measure real growth in assets. Prices of all types of asset moved up. We have therefore to deflate the 1974-75 value for price rise. These movements were unequal, their pace differed for different assets. There is no satisfactory deflator available which would take care of these developments. Since arbitrariness is involved in all index numbers, and since detailed data to build index are not available, a simple deflator is used by us. In Matar, paddy is a predominant crop, income and hence investment effort would be related to paddy production and prices. We have therefore used paddy prices as a deflator. We however do not claim it to be a satisfactory measure.

Table 11.1 gives details of values of different types of assets. Breakdowns are given for different size groups of landholders.

The aggregate money value of all assets, including land increased by 156 per cent (Table 11.1). Deflated value of assets indicates an increase of 53 per cent which gives an approximation of real increase in the value of assets in 1974-75 over 1965-66. When we exclude land from the assets, the deflated value of non-land assets increased by 42.2 per cent. Individually, increase in real value of land seems to be relatively much greater. All of it is not the result of investment effort in terms of consumption (current or future) foregone. For all non-land

TABLE 11.1: Value of Assets in Matar Taluka in 1965-66 and 1974-75

(Value in Rupees)

Asset Types	Size Class of Holding (Acres)					
	0 - 5		5 - 10		10 - 15	
	1955-66	1974-75	1965-66	1974-75	1965-66	1974-75
1	2	3	4	5	6	7
Land	5,65,206	27,75,259	11,38,020	43;19,811	10,43,942	25,34,095
House	3,89,928	15,28,369	6,63,650	18,89,412	8,57,524	10,11,830
Agricultural implements	57,156	2,90,331	1,14,503	3,92,784	1,93,875	3,17,516
Wells and Lifting devices	55,886	1,92,374	77,114	2,93,171	74,567	3,54,693
Livestock	1,27,012	5,44,076	1,49,555	5,14,263	1,14,337	2,80,,840
Liquid and Semi-liquid assets	74,937	5,70,631	1,93,954	6,89,194	2,01,332	5,25,006
All Assets	12,70,125	59,01,040	23,36,796	80,98,635	24,85,577	50,23,980
Number of Families	225	272	154	177	79	76
Percentage change in (nominal) value	364.60		246.57		102.13	
Percentage change in (real) value in terms of paddy equivalent	178.76		107.94		21.27	
Percentage change in (Nominal) value excluding land	343.42		315.22		70.54	
Percentage change in (real) value in terms of Paddy equivalent (excluding land)	266.05		189.13		42.33	

Contd.

TABLE 11.1: *(Contd.)*

(Value in Rupees)

Asset Types	Size Class of Holding (Acres)					
	15 - 25		25 +		All Cultivators	
	1965-66	1974-75	1965-66	1974-75	1965-66	1974-75
1	8	9	10	11	12	13
Land	8,85,347	15,72,889	11,91,647	20,17,887	48,24,162	1,32,19,941
House	4,73,991	4,15,769	4,67,769	6,18,685	28,52,862	54,64,065
Agricultural implements	72,791	1,40,683	1,18,530	3,24,038	5,56,855	14,65,352
Wells and Lifting devices	60,603	1,11,039	69,847	3,12,505	3,38,017	12,63,782
Livestock	66,020	1,02,749	86,781	1,14,213	5,43,705	15,56,141
Liquid and Semi-liquid assets	1,34,072	1,69,071	1,82,028	3,32,966	7,86,323	22,86,868
All Assets	16,92,824	25,12,200	21,16,602	37,20,294	99,01,924	2,52,56,149
Number of Households	43	24	34	29	535	578
Percentage change in (nominal) value	48.40		75.77		156.06	
Percentage change in (real) value in terms of paddy equivalent	-10.96		5.46		53.04	
Percentage change in value (Nominal) excluding land	16.33		84.05		137.03	
Percentage change in (real) value (in terms of Paddy equivalent) excluding land	-30.20		10.43		42.22	

Source: Field Survey, 1965-66.
Field Survey, 1974-75.

assets including housing too, non-inflationary component may not be all qualitative improvement. But it would still represent consumption foregone or real investment of part of income earned. Growth of all non-land assets, through net investment, at an annual compound rate of 11 per cent could be regarded rapid.

Table 11.2 gives growth rates of output and capital (all assets) including land. While real value of capital increased at an annual compound rate of 4.9 per cent, crop output increased at the rate of 3.1 per cent. At current prices the two increased at the rate of 11.0 and 9.1 per cent respectively. The average annual growth rate of gross value of agricultural produce and value of assets at current prices were nearly the same for farmers in size class with less than 5 acres, they diverged as the size of holding increased. Both the rates declined, for assets from 18.6 to 6.48 (4.5 for 15-25 acres group) and for output from 18.5 to 1.75 (0.65 for 15-25 acres groups). Such decline obtained in the growth rate of capital, despite a trend towards mechanisation that required lumpy investments. When we look at the growth in assets between 1965-66 and 1974-75, we find that the growth in quality of assets was higher for the lower size classes, the rate declined steeply as the size increased. This is a significant observation and can be attributed to public irrigation expansion to a large extent (Table 11.3).

TABLE 11.2: Growth Rates of Output and Capital (1965-66 to 1974-75)

(Per cent)

Size of holding (acres)	Average Annual Compound Growth Rates of			
	Gross value of agricultural produce		Value of all assets	
	At current prices	In terms of paddy equivalent	At current prices	In terms of paddy equivalent
1	2	3	4	5
0 - 5	18.5	11.9	18.6	12.0
5 - 10	13.7	7.5	14.8	8.5
10 - 15	6.9	1.0	8.2	2.1
15 - 25	0.7	-3.5	4.5	-1.2
25 +	1.8	-3.0	6.5	0.6
All Cultivators	9.1	3.1	11.0	4.9

TABLE 11.3: Value of Assets for Cultivator Households (1974-75 over 1965-66) (1965-66=100)

Size of holding (acres)	Change in assets value (current prices)			Change in asset value (in terms of paddy equivalent)		
	A	B	C	A	B	C
1	2	3	4	5	6	7
0 to 5	364.6	343.5	407.1	178.7	166.1	204.2
5 to 10	246.7	211.8	245.6	107.9	87.1	107.4
10 to 15	102.1	71.9	150.9	21.3	3.2	50.6
15 to 25	48.4	16.2	57.0	-11.0	-30.2	-5.8
25 +	75.8	83.7	137.0	5.5	10.22	42.2
All Cultivators	155.1	136.0	193.2	53.0	41.6	75.9

A = All assets.
B = Assets excluding land.
C = Assets excluding land and houses (Agricultural implements, wells and lifting devices, livestock, and liquid and semi-liquid assets).
Source: Field Survey, 1974-75.

Assets Structure

Widening base of capital is not accompanied by any major structural change (Table 11.4). Share of land in total assets remained around 52 per cent. Similarly liquid and semi-liquid assets accounted for around 9 per cent both in 1965-66 and in 1974-75. A shift did occur between housing and production assets, share of houses declining from 27 to 22 per cent and share of implements, water lifting devices and livestock increasing by about 2 percentage points each. The share of the value of the land increased for all size classes except the top, 25+ acres class, share of the value of houses declined for all size classes, decline being distinctly large for 10 + acre size classes. Livestock showed an erratic trend giving marginal increases or decreases. Share of water-lifting devices increased sharply for size classes of 10+ acres and of implements for 15+ acre size classes. None of these changes is marked. Nevertheless, these changes are indicative of increased mechanisation, in the case of relatively bigger farmers.

Asset Distribution

Table 11.5 gives gini ratio of asset distribution across size class of

TABLE 11.4: Composition of Assets of Cultivator Households in Matar Taluka, 1965-66 — 1974-75

(Per cent)

Assets	Size of holding (in acres)											
	0 - 5		5 - 10		10 - 15		15 - 25		25 and above		All Cultivators	
	1965-66	1974-75	1965-66	1974-75	1965-66	1974-75	1965-66	1974-75	1965-66	1974-75	1965-66	1974-75
1	2	3	4	5	6	7	8	9	10	11	12	13
Value of land	44.5	47.0	48.7	53.3	42.0	50.4	52.3	62.6	56.3	54.2	51.9	52.3
Value of houses	30.7	25.9	28.4	23.3	34.5	20.1	28.0	16.6	22.1	16.6	27.1	21.6
Value of agricultural and other implements	4.5	4.9	4.9	4.9	7.8	6.3	4.3	5.6	5.6	8.7	5.6	5.8
Value of water lifting devices and wells	4.2	3.3	3.0	3.6	2.9	7.1	3.6	4.4	3.2	8.4	3.2	5.0
Value of livestock including bullocks	10.0	9.2	6.4	6.4	4.6	5.6	3.9	4.1	4.1	3.1	4.6	6.2
Value of liquid and semi-liquid assets	5.9	9.7	8.3	8.5	8.1	10.5	7.9	6.7	8.6	9.0	8.4	9.1
Total value of assets per family including land Rs	5,645	21,695	15,174	45,755	31,463	66,105	39,368	1,04,675	62,253	1,28,286	18,509	43,702
Total value of assets per family excluding land Rs	3,133	11,492	7,872	21,357	18,334	32,764	18,801	39,137	27,255	58,706	9,534	20,826

Source: Field Survey, 1965-66.
Field Survey, 1974-75.

TABLE 11.5: Concentration of Assets and Production

	Gini ratios	
	1965-66	1974-75
Distribution of operational holding (Area)	0.46	0.42
Distribution of operational holding (value)	0.38	0.33
Distribution of all assets (value)	0.43	0.33
Distribution of all assets (excluding land) (value)	0.43	0.38
Distribution of all assets (excluding land and house) (value)*	0.40	0.33
Distribution of gross value of agricultural produce	0.46	0.27

* The assets in this category include water-lifting devices, wells, other agricultural implements, bullocks and other animals and liquid and semi-liquid assets.

holding for Matar taluka as a whole for 1965-66 and 1974-75. The comparison brings out a distinct trend of narrowing of inequality. We compare these ratios with that of gross value of output. A general trend of narrowing of inequality of asset distribution is observed irrespective of whether we consider all assets, assets without houses or without land and houses both. We have already observed unequal investment efforts, in terms of growth of assets for different size classes, the rate of growth of assets being inversely related to the size of holding. It would reflect, given expanding opportunities for investment, even in non-land assets, that small farmers made serious efforts to expand their enterprise, much of which could be 'bootstrap' growth. As we shall see later, compared to rise in assets, liabilities increased less rapidly.

Comparison of distribution of assets, with that of crop output is interesting. In 1965-66 gross value of crop output had gini coefficient of 0.46, which was higher than gini coefficients of assets. In 1974-75, the gini coefficient of crop output declined to 0.27, a substantial decline indeed. Gini values of assets distribution also declined but decline was relatively small. Between 1965-66 and 1974-75, irrigation expanded, green revolution was ushered in, crop pattern changed and use of fertilisers spread rapidly. All these changes seem to be not only size-neutral but perhaps of advantage to small farmers, who seem to have expanded their production to a much larger extent.

It is legitimate to argue that the narrowing of inequalities in assets and production distribution can be traced to probable subdivision of large holdings. We have shown in Tables 3.6 to 3.8 (Chapter 3) that families that did not branch out were much larger in number 746 (364 cultivators) compared to those that branched out 114 (76 cultivators).

The satellites themselves numbered 187 (83 cultivators). Even activity of this magnitude could affect land distribution, which showed a decline in Gini-ratio from 0.46 to 0.42.

But subdivision by itself may not give further advantage to subdivided households regarding quality of land for which we find a greater decline in inequality than that for area. One may concede that prior to or subsequent to subdivision, households affected by subdivision may invest in land and other assets. It is also possible that subdivision may be unequal over regions, one with higher land value bearing greater impact. In other words the improved distribution may be mainly, if not solely, a direct consequence of rising pressure of population and consequent deterioration.

We would view the process in a slightly different light. We would see the relatively much greater decline in Gini-ratio for all assets together from 0.43 to 0.33, a product of enthusiasm of small and medium farmers generated by new opportunity for investment as a result of expanding irrigation facilities, together with an effect of population growth the former a positive effect, the latter a negative effect, a net result being more desirable.

Secondly, we would also like to highlight the decline in gini-ratio of production distribution from 0.46 to 0.27. Benefits of new irrigation facilities would seem to be availed better and more widely by small and medium farmers in 1974-75. We recall our findings of Chapter 4, we found that small and medium size farmers had a greater degree of conjunctive use of irrigation, with an emphasis on well irrigation. We have shown further in chapters 5 and 7 that small farmers had higher cropping intensity and larger dose of fertilisers (Tables 5.4 and 7.3). It is possible that households newly entering lower size class of land holdings may have greater motivation to grow more produce, but their efforts alone would not enable the class of small and medium size farmers to turn the scales in the favour of the class they now belong. Besides, the population growth on Matar has been at nearly uniform rate since 1951, the phenomenon of subdivision is therefore not entirely new. If so, we would like to emphasise a switch in the relative position of inequality of distribution of assets and production, the latter being relatively very low in 1974-75 compared to that of assets, unlike the relative positions of the two obtained in 1965-66. All these developments read together cannot be considered mainly a product of deteriorating economic conditions consequent of rising pressure of population.

Investment in Physical Assets

We have examined growth of assets and distribution of investment through inventory method. We have some data of direct investment which gives useful information in terms of the extent of participation in investment by different groups of cultivator households. Table 11.6 gives percentages of farm families investing in physical assets during the last ten years. 210 out of 578 or 36 per cent of farm families invested in assets. This is indeed a low figure. An average life of most of the assets, excluding wells, houses and land, is about 8 to 10 years. In terms of gross investment – or replacement of old assets, all families would have incurred expenditure to purchase assets between 1965-66 and 1974-75. One-third of the total families have reported investment over the period of 9 years, most of them would have reported additional acquisition and not repairs and replacement. Even then participation is low. The economy is expanding under the impact of irrigation, rate of growth of assets was high, and yet majority of the cultivator households have not participated. Impact of irrigation on investment is thus not widespread. Yet asset distribution has become less unequal, mainly because participation among medium and small farmers is wider. Among large families relatively fewer farm families have participated in investment.

TABLE 11.6: Cultivator Household Investing in Physical Assets During 1965-66 — 1974-75 Period

(Per cent)

Size of holding (acres)	Matar Taluka
0 to 5	29
5 to 10	47
10 to 15	42
15 to 25	33
25+	28
All Cultivators	36
Total number of cultivator households	578
Cultivator households reporting investment	210

Source: Field Survey, 1974-75.

Debts : Levels and Uses

Agriculture is an individual enterprise, organised around family. Equity of risk capital is supplied therefore by the family. In a traditional economy family was required to borrow to tide over the period of scarcity. More often, debts were incurred to meet household maintenance expenses during production season. Borrowing for production activity is the feature of a growing economy, or economy linked with market mainly through cash crops. Term-borrowing is essentially a feature of expanding economy such as that of Matar between 1965-66 and 1974-75. Per farm family outstanding debt increased from Rs. 1,181 in 1965-66 to Rs. 1,974 in 1974-75, an increase of nearly 67 per cent. Since assets value also increased, the debt liabilities as percentage of total assets value declined from 6.30 in 1965-66 to 4.52 in 1974-75. Even as percentage of assets value excluding land the decline was from 12.4 to 9.5 (Tables 11.7 and 11.8).

Decline in debt burden was experienced by all classes of farmers, except those in size class of 15-25 acres. Those in size class of 10-15 acres experienced an increase in percentage of debt to value of assets excluding land. The decline was substantial in the case of both small and big farmers, middle size farmers experienced either small decline or substantial increase. This class of farmers participated in investment to a much greater extent.

The burden of debt is measured several ways. For farmers, land is a major asset, debt per acre is therefore commonly accepted as a measure of debt burden. Since land held by cultivator household in the sample declined and since debt in absolute amount increased, per acre debt increased more steeply from Rs. 82 in 1965-66 to Rs.281 in 1974-75, increase being 243 per cent. But value of land increased still more sharply. Per acre land value went up from Rs. 1,020 to Rs. 3,269 between 1966 and 1974-75. Consequently, debt relative to value of land declined from Rs. 16.8 to Rs. 8.6 per Rs. 100 worth of land or by nearly 49.9 per cent. This is a substantial reduction. Debt measured in terms of burden on output also declined. Per Rs. 100 worth of output amount of outstanding debts declined from Rs. 36.1 in 1965-66 to Rs. 28.5 in 1974-75 or by 21 per cent.

Debt includes outstanding amount of not only term loans but also accumulated burden of current or short and medium term borrowings. In a traditional economy debt once incurred used to multiply since repaying capacity was low. The fact that outstanding debt worked out to be about 28.5 per cent of the current value of outputs the burden may not be considered low, if it represented all accumulated burden of past short-term borrowings. Otherwise a decline in percentage of debt to value of output would represent an improvement. Total investment made over 9 years, represented about 30 per cent of initial capital, but

TABLE 11.7: Outstanding Debts of Cultivator Households, 1965-66 and 1974-75

Size of holding (acres)	Debt per family (Rs.)		Debt per acre of land (Rs.)		Debt per Rs.100 worth or land (Rs.)		Debt per Rs.100 worth of gross produce of crops (Rs.)	
	1965-66	1974-75	1965-66	1974-75	1965-66	1974-75	1965-66	1974-75
0 - 5	563	1125	197	442	22.2	11.0	45.5	30.4
5 - 10	1026	1610	137	222	14.5	6.6	33.5	22.9
10 - 15	1630	3346	133	297	12.4	10.0	27.3	33.1
15 - 25	1714	6563	94	375	8.8	10.0	22.6	47.7
25 +	4234	4755	119	172	12.1	6.8	23.7	23.4
All cultivators	1181	1974	82	281	16.8	8.6	36.1	28.5

Sources: Field Survey, 1965-66.
Field Survey, 1974-75.

TABLE 11.8: Debt as Percentage of Total Asset Value per Farm Family (1965-66 and 1974-75)

Size of holding (acres)	Debt as percentage of value of all assets		Debt as percentage of value of assets (excluding land)	
	1965-66	1974-75	1965-66	1974-75
1	2	3	4	5
0 to 5	10.0	5.2	18.0	9.8
5 to 10	6.8	3.4	13.0	7.5
10 to 15	5.2	5.1	8.9	10.2
15 to 25	4.4	6.3	9.1	16.8
25 +	6.8	3.7	15.5	8.1
All Cultivators	6.4	4.5	12.4	9.5

Sources: Field Survey, 1965-66.
Field Survey, 1974-75.

for the current year, it would represent approximately 15 per cent of current value of output. Burden of debt would be greater than this. But, debt represented partly borrowing for long term investment, the repayment for which will spread over subsequent years. Even from this angle declining debt burden could imply greater repayment than fresh borrowings.

Both the sources of borrowings and purpose for which debts were incurred have undergone structural changes. These changes are significant from the point of burden of debt. An overwhelming feature of borrowing in 1965-66 was that about 63 per cent of borrowing was for meeting current expenses. After irrigation expansion, the need for current production expenses (circulating capital) steeply increased. However current expenses continued to have significant share which was 45.3 per cent of the total outstanding debt in 1974-75 (Table 11.9). The major change is represented by a steep increase in the borrowing for farm investment. Its share in total outstanding debts, increased from 3.8 per cent in 1965-66 to 19.2 per cent in 1974-75.

Share of household maintenance expenditure and of expenditure for social ceremonies remained nearly unchanged. Thus the productive and unproductive debts increased in the same proportion. But productive expenses underwent a change, shift being in favour of long-term investments. The share of borrowing for different purposes changed for all size classes of land holdings in the same direction. However, for small farmers the non-productive uses still dominated, for medium and big farmers the share of investment steeply increased.

TABLE 11.9: Distribution of Borrowings of Farm Families According to Purpose 1965-66 — 1974-75 (By size Class)

Size of holding (acres)	Year	For Current farming expenses	For Farm investment	For family consumption	For celebrating social events	Others	Total
1	2	3	4	5	6	7	8
0 - 5	1965-66	38.2	3.3	29.5	27.7	1.3	100.0
	1974-75	29.4	8.3	33.3	22.2	6.8	100.0
5 - 10	1965-66	52.0	3.4	17.4	22.6	4.6	100.0
	1974-75	58.6	12.6	7.3	13.6	7.9	100.0
10 - 15	1965-66	73.3	4.3	14.9	2.7	4.8	100.0
	1974-75	46.7	13.5	6.7	28.9	4.2	100.0
15 - 25	1965-66	75.9	11.7	2.4	10.0	—	100.0
	1974-75	41.3	47.6	—	—	—	100.0
25 +	1965-66	81.6	—	2.0	9.3	7.1	100.0
	1974-75	54.7	34.9	10.4	—	—	100.0
All Cultivators	1965-66	63.1	3.8	14.1	15.0	4.0	100.0
	1974-75	45.3	19.2	13.5	15.8	6.2	100.0

Sources: Field Survey, 1965-66.
Field Survey, 1974-75.

Burden of debt depends on interest liabilities. Borrowings from private agencies carry relatively greater interest liabilities, institutional loans carry relatively low interest burden. Between 1965-66 and 1974-75 share of loans from moneylenders sharply declined, from 26.1 per cent to 19.2 per cent. Share of co-operative loans increased from 34.3 per cent in 1965-66 to 53.4 per cent in 1974-75. Land Development Banks and Commercial Banks together accounted for 15.5 per cent in 1965-66 and 22.2 per cent in 1974-75. Personal loans from relatives accounted for 23 per cent in 1965-66, in 1974-75 along with the rest, these loans accounted for only 5.1 per cent (Table 11.10). Share of institutional agencies increased from 51 per cent to 76 per cent. Decline in the share of non-institutional agencies and increase in the share of institutional agencies was much sharper in the case of small farms. For small farms, though share of institutional agencies increased, the increase was confined to short-term credit.

TABLE 11.10: Sources of Borrowings by Cultivator Households 1965-66 and 1974-75

(*Per cent*)

Size	Agencies of lending, 1965-66					Agencies of lending, 1974-75					
	Money-lenders	Government	Co-operative societies	Land Development Bank	Relatives	Money-lenders	Government	Co-operative societies	Land Development Bank	Other Banks	Others
1	2	3	4	5	6	7	8	9	10	11	12
0 - 5	44.1	0.6	20.4	7.7	27.2	20.4	0.6	60.8	3.6	3.9	10.7
5 - 10	27.3	2.0	36.3	10.5	23.9	24.3	0.2	61.6	5.7	2.5	5.6
10 - 15	20.9	0.5	37.5	25.3	15.8	20.4	0.0	39.3	35.3	1.4	3.5
15 - 25	19.7	2.7	53.1	18.7	5.8	2.5	0.6	38.1	58.7	0.1	0.0
25 +	16.7	1.9	30.9	15.5	22.6	22.3	0.0	63.2	14.5	0.0	0.0
All Cultivators	26.1	1.5	34.3	15.5	22.6	19.2	0.3	53.4	20.1	2.0	5.1

Sources: Field Survey, 1965-66.
Field Survey, 1974-75.

Summing-up

Public sector investment in irrigation seems to have encouraged investment in the private sector, the rate of growth is fairly high. A high rate of investment has been accompanied by reduced inequality of distribution of assets. Growth of capital however has raised capital intensity, which was accompanied by rapid growth of crop production and narrowing of inequality of crop production.

The overall growth of assets is accompanied by reduced relative burden of debt despite nearly doubling of absolute amount of debt per family and over two and half time increase in per capita debt. Burden of debt seems to have decreased, even in terms of relative burden of interest liabilities.

The structure of both, assets and debts, changed in right direction. Non-land physical assets expanded more rapidly. Share of institutional agencies in outstanding debt increased significantly.

Reduced burden of debt along with increased assets implied building up of assets to a greater extent from own savings. An observation of major economic significance emerging from Matar experience is that public sector large-scale investment would encourage rapid expansion of investment in private sector and high rate of investment might lead to reduced inequality. We shall examine this particular phenomenon through analysis of consumption data as well.

An unwelcome feature that would cause concern pertains to participation in capital formation. The investment participation remains low. However it was wider in Navagam relative to other two regions. In this one can see a hope for future for widening participation when irrigation becomes equally mature in all the three regions.

12

LEVELS AND PATTERN OF CONSUMPTION AND THE EXTENT OF POVERTY

IN THIS chapter we examine effects of irrigation on personal consumption. When income rises, level of total consumption expenditure rises but expenditure on individual items of consumption rises at unequal rates. In the context of consumption it is therefore interesting to know how in an expanding economy of a region shares of expenditure on important groups of items (as well as on individual items) change. Shifts in shares are results of the price and income elasticities of demand for commodities. We examine if elasticities have also changed between 1965-66 and 1974-75.

In a growing economy it is often found that incomes and therefore expenditure of all classes of the society do not experience same or even similar change. Growing inequality often accompanies rising levels of income and consumption at the level of a region or a nation. Whether such a shift in incomes also involves an absolute deprivation of a class of people is of immediate relevance. The former, i.e., income inequality, and the latter, i.e., absolute deprivation are referred as relative and absolute poverty. A special concern is shown for the latter. We examine both the relative deprivation and the absolute poverty levels for years 1965-66 and 1974-75.

We remind ourselves that our data refer to two points of time, 1965-66 and 1974-75. Agricultural incomes fluctuate as a consequence of weather as well as price fluctuations. Irrigation partly protects production from weather uncertainties. In our surveys both 1965-66 and 1974-75 years were below normal, the rainfall in both the years was deficient. During the years of deficient rainfall farmers suffer more compared to families in other occupations. Consumption is influenced not only by current incomes but also by incomes of previous years. Since income levels of farmers do not rise or fall monotonically, influence of income levels of previous years is also not uniform for all years. We shall see such uneven influence in our analysis later.

Levels of Consumption

We begin with changes between 1965-66 and 1974-75 in absolute levels of overall consumption expenditure for different occupations and also by consumption decile groups. We should note that total consumption expenditure includes only current expenditure, we have not taken account of expenditure on consumer durables and expenditure on social and religious functions. These do not recur with a uniform pattern from one year to the next.

We give consumption data in Tables 12.1 and 12.2. They give per capita current consumption expenditure levels (in terms of current prices) for 1965-66 and 1974-75. For comparing changes in consumption levels over time, consumption expenditure given at current prices, need to be adjusted for price changes by an appropriate price deflator. For Matar we have not evolved one. We have two options open to us. We have for rural Gujarat cost of living index for agricultural labour. For Matar we do not have a similar index. Considering all classes together we contend that, income levels of the region would be influenced importantly by production and prices of paddy since paddy is a major crop. If this contention is accepted, paddy price index can be taken as a deflator. In a cost of living index we allow prices of all items to vary but keep their relative weights the same, in the case of single article price we assume even equal movement of prices of all relevant commodities. The latter would not be a major vitiating factor, if the item selected has an overwhelming by large weight in the index considered. In gross value of farm output value of paddy accounted for close to 60 per cent. Income levels of farming section, and that of others, whose income would depend on farm economy, would be influenced predominantly by paddy prices. On consumption side share of rice in 1974-75 was only around 15 per cent of the total expenditure in rural Gujarat. We have used both paddy prices and cost of living index for agricultural labourers in rural Gujarat as alternative price deflators. The trend differs according to the deflator used. The results are interesting. Taking all families together and using paddy price deflator (PPD), we fiind the real average per capita expenditure (APCE) to have increased by about 19 per cent. This is a modest increase considering rapid expansion of irrigated area.

Inter occupation groups' comparison of the changes in APCE provides an explanation for such modest increase. Between farm and non-farm families, the increase in the real APCE was greater for non-farm families. It was 39 per cent for non-farm families as against 10 per cent for farm families. We expect a major expenditure increase for farm families as they would have benefited directly from irrigation. Year 1974-75 was below average, yields of crops were adversely affected. However we expect expenditure to reflect a permanent income position in which current income would have relatively lower weight especially for the upper income groups with backing of accumulated assets. As the data of APCE suggest, there is relatively much smaller change in the importance of the current income level in influencing a given year's expenditure level. Examining the changes in APCE for size groups within the class of farm families, we observe the impact of weather more clearly. For instance, the highest increase in real APCE was reported among small farmers (0-5 acre). As the size of holding increased the increase in real APCE declined. Decline in real APCE was also observed for households with holdings between 10-15 acres and those with above 25 acres. We have already observed that the extent of area under irrigation tended to decline as the size of holding increased. The extent of irrigated area increased for all size classes, but a severely deficient year like that of 1974-75, seems to have wiped out the gain of irrigation at least for big and medium farmers. Agricultural labourers as a group had the largest increase in real APCE, of about 93 per cent. While output fluctuates widely under the influence of weather, fluctuation in inputs including labour would be much smaller. The general increase in the labour use and their reward, subsequent to irrigation expansion is more clearly reflected in the APCE of farm labour.

Among non-farm families the increase in real expenditure varied between 26 per cent for salary earners and 55 per cent for 'artisans', and 'traders' and 'others' had nearly 50 per cent increase, which is close to that of small farmers. Instead of paddy price deflator, if we use cost of living index for agricultural labourers as deflator most of the above conclusions change. APCE at the level of all households declined by about 15 per cent. All farm households except small farmers (0 to 5 acres) experienced a decline in per capita expenditure. We should recall that population growth in Matar taluka between 1965-66 and 1974-75 was around 3 per cent per annum (compound rate). Large part of the gain of irrigation was eroded by population catching up with rising household income.

TABLE 12.1: Per Capita per Annum Expenditure by Occupations, 1965-66 and 1974-75

(Rupees)

Occupation Group	Per Capita Expenditure Per Annum									
	No. of households	1965-66	Rank	No. of households	1974-75 (Current prices)	Rank	1974-75 Constant PPD Prices 1965-66 = 100 1974-75 = 167	1974-75 Constant C OLI Prices 1965-66 = 100 1974-75 = 235	Change with PPD per-cent	Change with COLI Per cent
1	2	3	4	5	6	7	8	9	10	11
I. *Cultivators*										
Size of Holding (acres)										
0 - 5	225	273.50	9	272	711.39	9	425.98	303.11	55.75	10.83
5 - 10	151	392.92	6	177	720.30	8	431.36	306.94	9.78	- 21.88
10 - 15	82	497.07	3	76	746.93	6	447.26	318.25	- 10.02	- 36.97
15 - 25	43	503.13	2	24	838.21	5	501.92	357.14	- 8.24	- 29.02
25 +	34	660.63	1	29	947.74	2	567.51	403.81	- 10.10	- 22.02
All Cultivators	535	402.75		578	737.08		441.36	314.05	9.59	- 22.02
II. Agricultural Labourers	244	215.48	10	314	693.21	10	415.09	295.36	92.63	37.07
III. *Non-Farm*										
Artisan	35	313.54	8	38	745.55	7	446.44	317.66	42.39	1.31
Trader	41	396.66	5	43	995.71	1	596.23	424.25	50.31	6.95
Service	79	403.49	4	75	850.60	4	509.34	362.42	26.23	- 10.17
Others	64	331.17	7	63	859.11	3	514.44	366.55	55.34	10.68
All non-farm	219	369.54		219	857.89		513.71	365.53	39.01	- 1.08
IV. All Households	998	373.00		1111	743.37		445.13	316.73	19.34	- 15.08

Note: PPD = Paddy Price Deflator
COLI = Cost of Living Index for Farm Labour

Source: Field Survey, 1974-75.

TABLE 12.2: Per Capita Monthly Expenditure in by Deciles (1965-66 - 1974-75)

(Rupees)

Decile Group	Mean expenditure levels					
	1965-66	1974-75			Per cent Change	
	Current Price	Current Price	Deflated on		PPD	COLI
			PPD	COLI		
1	2	3	4	5	6	7
0 - 10	10.16	27.33	16.36	11.63	61.0	14.5
10 - 20	14.26	36.18	21.66	15.39	51.9	7.9
20 - 30	16.84	41.88	25.08	17.82	48.9	5.8
30 - 40	19.16	46.56	27.88	19.81	45.5	3.4
40 - 50	21.55	51.35	30.75	21.85	42.7	1.4
50 - 60	24.79	57.20	34.25	24.34	38.2	- 1.8
60 - 70	28.46	63.93	38.28	27.20	34.5	- 4.4
70 - 80	34.47	72.00	43.11	30.64	25.1	- 11.1
80 - 90	43.42	83.96	50.27	35.73	15.8	- 17.7
90 - 100	75.56	137.19	82.15	58.38	8.7	- 22.7
All households	31.08	61.94	37.08	26.36	19.3	- 15.1

PPD = Paddy Price Deflator.
COLI = Cost of Living Index for farm labour (Rural Gujarat).
PPD = 1965-66 = 100 1974-75 = 167.
COLI = 1965-66 = 100 1974-75 = 235.
Source: Field Survey, 1974-75.

The unequal increase in APCE for different classes seem to narrow inter-occupation group inequality. We shall examine changes in expenditure inequality later. It is interesting to observe at this stage that in regard to APCE the agricultural labour household ranked 10th – the lowest, both in 1965-66 and 1974-75. Low rank of agricultural labourers on expenditure level obtained despite the fact that they realised highest increase in real APCE. Similarly, the small farmers were just above agricultural labourers, with 9th rank in both the years. At the top the large farmers with operational holding of 25 acres and above ranked 1st in 1965-66, and in 1974-75 they ranked next to trading households, the latter ranked first. The non-agricultural families improved their ranks *vis-a-vis* farm families, the latter perhaps being adversely affected by weather. The rank correlation for 1965-66 and 1974-75 APCE was 0.677.

We may examine the change in APCE as reflected by expenditure decile groups. A clear trend in the increase in APCE with paddy price deflator emerges. The increase in APCE varied between 9 and 61 per cent. We find the highest increase was 61 per cent, and the percentage in expenditure increase monotonically declined as one moved up the decile groups. While comparing the decile groups at two points of time it must be mentioned that the same families are not necessarily in the same decile group at two points of time.

We may therefore find that those at the top in 1965-66, could appear in 1974-75 among lower deciles. The decile transition is a useful way of looking at the process of change and the families involved in the change. Pending that we may measure the change in inequality of the APCE among deciles at two points of time. Value of gini coefficient of APCE for all families was 0.33 in 1965-66, it came down further to 0.25, a significant change. We discuss, changes in inequality, along with the changes in the welfare and the poverty levels later in greater details.

Changes in Expenditure Levels and Budget Shares

Between 1965-66 and 1974-75, the total consumption basket improved, its size increased, as well as its contents improved for most, if not for all occupation groups. We now inquire whether all its contents, or the selected ones showed increase and a few showed reduction. We examine the changes in terms of constant prices (current prices deflated by using PPD). With the help of Table 12.3 we observe the following:

(1) for small farmers (a) all items showed increase in expenditure except expenses on travelling, (b) percentage increase was greater for food items than non-food items, (c) in food, the lowest increase was in regard to foodgrains (19.5%), the highest was in regard to meat/eggs (286%) followed by edible oil, milk, vegetables and sugar, tea in that order, (d) for non-food items the largest increase (479.49%) was for entertainment and (432.46 per cent) for toiletry and the smallest (1.8%) in tobacco, (e) expenditure on clothing, education and health increased by 88.55, and 32 per cent respectively, and (f) between food and non-food, the former increased twice as fast the latter.

(2) for farmers with 5 to 10 acres holding (a) milk, sugar and tea

registered a decline; among food expenditure on foodgrains was nearly unchanged and that on meat, vegetables and oils increased in that order, (b) between food and non-food the former increased rapidly, but the overall increase was small (11%), (c) among non-food items travelling, health, education and tobacco registered a decline. The highest increase (980 per cent) was for reading material followed by entertainment (470%) and soap and toiletry (330%).

(3) for farmers in size class of 10 to 15 acres, there was a 10 per cent decline in APCE, decline being around 9 per cent in expenditure on food and 13 per cent in non-food items, (b) expenditure on edible oil, sugar, tea, and vegetables among food items and on books, toiletry, entertainment increased followed by that on education.

(4) for the next two size groups 15 to 25 acres and 25 acres and above, APCE was constant or had declined (a) in both the cases expenditure on food declined, that on non-food increased, (b) among food, expenditure on foodgrains and sugar and tea declined, on other food items expenditure increased for 15-25 acres group, but declined for big farmers, meat being exception, among non-food items expenditure on education, reading material-declined for both size classes, for both the classes the highest increase was for entertainment and soap and toiletry.

(5) we have already observed that the agricultural labour experienced a fairly large increase in total expenditure. The increase was larger for non-food than that for food items and except a catch-all item of other non-food all other items experienced increase. Among food, the largest increase was for meat followed by edible oil, vegetables, sugar, tea and milk and other milk products in that order.

A striking feature was a decline in the percentage share of expenditure on foodgrains for cultivators and labourers (Table 12.4), the greatest decline being in the case of labour families. Similarly, decline was observed in the share of expenditure on milk and milk products, decline being the largest for non-farm families, share of expenditure on sugar, tea etc. increased for all the three broad occupation groups. A shift away from own product and in favour of those purchased from the market even in a lean year would be indicative of a long term shift in preferences. Within the group of

TABLE 12.3: Change in per Capita Annual Expenditure (Real) by Occupation Groups and Commodity Groups (1965-66 to 1974-75)

(Percentage change in real expenditure in terms of kilograms of paddy equivalent)

Commodity Group	Cultivators						Agricultural Labourers	Non-farm households					All households
	0-5	5-10	0-15	15-25	25+	All Cultivators		Artisans	Traders	Salary earners	Others	Non-farm households	
1	2	3	4	5	6	7	8	9	10	11	12	13	14
Foodgrains	19.5	0.5	18.9	13.4	26.1	7.6	55.0	22.94	76.6	3.94	81.7	40.7	11.8
Milk and other Milk Products	68.2	- 2.3	- 12.9	- 11.2	- 16.4	3.7	79.3	8.1	12.0	20.2	- 8.5	7.8	9.5
Edible oil	119.9	54.3	6.0	16.5	- 38.3	36.7	163.0	128.5	88.3	61.9	77.6	85.7	72.9
Sugar, tea etc.	59.5	- 19.5	12.8	- 12.1	- 32.2	11.5	97.3	25.7	43.8	49.0	14.3	37.1	28.0
Vegetables	65.1	32.2	21.4	16.5	- 28.4	28.6	158.4	75.2	91.1	151.1	132.6	116.3	51.9
Meat, eggs	286.1	53.3	- 40.9	78.3	63.4	38.2	336.8	-	377.8	283.5	612.2	312.0	93.8
Total Food	149.1	11.9	- 8.6	0.6	- 22.9	10.5	91.3	39.9	69.6	35.0	66.0	50.7	28.1
Fuel and light	178.8	86.0	55.9	63.3	101.4	94.9	254.5	200.4	73.0	80.3	18.9	88.0	36.6
Clothing	87.8	47.6	6.8	28.9	15.0	34.2	89.1	56.7	37.7	41.0	40.2	39.1	24.6
Education	54.5	- 36.5	75.9	- 25.11	- 70.2	49.5	7.8	53.8	- 58.1	- 55.5	24.1	- 53.7	- 50.2
Medicine	31.6	- 52.3	- 51.1	- 45.8	158.7	- 31.6	126.7	49.9	73.2	- 3.2	18.1	10.6	- 16.8
Entertainment	1479.5	470.0	277.8	525.8	255.0	405.4	3267.5	886.7	184.8	127.0	699.4	229.8	411.2
Travelling	- 14.7	- 37.6	- 65.3	- 66.0	21.9	- 41.5	22.2	- 45.3	- 67.8	- 56.0	59.7	- 64.9	- 38.2
Tobacco	1.8	- 7.1	- 8.4	16.9	- 6.7	- 7.4	46.4	95.7	54.7	- 15.5	65.7	29.7	7.5
Soap and Toiletry	432.5	330.0	230.6	297.0	218.0	286.8	656.8	503.5	256.3	264.5	464.0	298.1	315.0
Books and Newspapers	-	980.0	920.0	- 10.0	- 10.5	425.0	-	325.0	- 13.2	245.5	14.1	116.7	238.8
Other non-food	13.2	- 68.9	- 73.1	- 56.8	- 12.2	- 19.9	- 15.6	- 44.2	- 23.5	- 30.3	- 80.6	- 37.8	- 57.2
Total non-food	78.3	5.4	- 12.5	- 1.0	5.2	8.2	137.0	48.9	19.7	12.6	55.7	18.0	9.6
Total Expenditure	56.1	10.0	- 9.8	- 0.0	- 13.7	9.8	98.1	42.7	50.6	26.5	55.7	39.3	22.4

Sources: Field Survey, 1965-66.
Field Survey, 1974-75.

non-food items, the shares of expenditure on clothing and fuel (which now includes changes in expenditure for consumption of electricity, kerosene to a larger extent) increased for farm families. A decline in the case of all remaining non-food items taken together in the case of farm and non-farm families, could be due to increased public expenditure on health care and education. In the case of agricultural labourers shares of expenditure on fuel went up. It could partly be due to curtailment of supply of fuel gathered free form the village common land. Free supply of cotton stumps for fuel to labourers by employers might also have dwindled as area under cotton declined. The labourers would be required to buy fuel. Dominance of intra-food and non-food groups shifts rather than that of shifts between groups of food and non-food items is a combined result of two opposite forces, (1) commercialisation of food consumption and (2) increased social expenditure for items of human capital. In the absence of the latter, substitution could have been of different type and the emerging pattern of budget shares would have been different. The normal expectations of shifts in budget shares along with the rising incomes are not observed under the circumstances. Nevertheless we do get an insight into overall improvement of standard of living in an observable manner.

A few specific instances of different occupation groups throw more light on the type of changes in the budget shares. Let us compare expenditure pattern of small (0-5 acres) and big (25+ acres) farmers. For both the groups experienced a decline in the share of expenditure on food articles. But, decline in the percentage share of expenditure on foodgrains is marked for small farmers and modest for big farmers. This is expected as big farmers spent relatively less on foodgrains in 1965-66. In fact small farmers' relative expenditure on foodgrains in terms of percentage in 1974-75 was as much as that of big farmers in 1965-66. What is striking is that in the case of small farmers decline in share of expenditure on foodgrains was matched to an extent by a sharp increase in the share of expenditure on 'other foods' and a marginal increase in the share of expenditure on milk and milk products, and less significant increase on sugar, tea etc. In the case of big farmers shares of expenditure on all items of food declined, shift in their case was in favour of non-food items, mainly clothing and fuel and lighting. Regarding non-food items, share of expenditure on all items experienced modest to marginal increases. Thus we find, as would be expected on the basis of the demand theory, that at the lowest

TABLE 12.4: Budget Shares in the Per Capita Expenditure by Occupation and Commodity Groups (1965-66 to 1974-75)

Commodity Groups	Cultivators land size (acres)											
	0 to 5		5 to 10		10 to 15		15 to 25		25 +		All Cultivators	
	1965-66	1974-75	1965-66	1974-75	1965-66	1974-75	1965-66	1974-75	1965-66	1974-75	1965-66	1974-75
1	2	3	4	5	6	7	8	9	10	11	12	13
Food	48.0	36.6	40.0	36.8	38.0	33.8	38.0	32.6	36.0	33.2	40.6	35.6
Milk and other Milk Products	9.3	10.0	12.3	11.0	11.8	11.4	12.2	13.5	12.0	11.7	11.4	10.1
Tea, Sugar etc.	9.1	9.3	8.0	8.4	7.8	7.6	7.3	6.4	7.9	8.9	8.2	8.3
Other food	10.2	17.3	10.2	21.6	10.0	15.8	9.0	13.6	12.1	9.3	10.4	17.0
Total food	76.6	73.2	70.4	77.8	67.7	68.7	56.5	66.2	68.1	63.0	70.6	71.0
Clothing	10.3	12.5	9.9	13.3	11.6	13.7	10.2	13.2	12.5	16.7	10.8	13.2
Fuel and lighting	1.4	2.6	1.7	2.8	1.9	3.4	1.6	2.5	1.5	4.0	1.6	2.9
Other non-food	11.7	11.8	17.8	12.1	18.7	14.3	21.8	18.1	17.9	18.3	17.0	12.9
Total non-food	23.4	26.8	29.6	22.2	32.3	32.4	33.5	33.9	31.9	37.0	29.4	29.0
Total	100.0	100.0	100.0	100.0	100.0	100.0	100.0	100.0	100.0	100.0	100.0	100.0

Contd.

TABLE 12.4 *Contd.*

Commodity Groups	Non-farm occupation groups										All non-farm households		All Households	
	Agricultural Labourers		Artisans		Traders		Salary Earners		Others					
	1965-66	1974-75	1965-66	1974-75	1965-66	1974-75	1965-66	1974-75	1965-66	1974-75	1965-66	1974 75	1965-66	1974-75
1	14	15	16	17	18	19	20	21	22	23	24	25	26	27
Foodgrains	49.6	39.8	37.3	33.4	37.2	36.2	37.5	27.3	38.0	43.0	34.4	34.7	39.7	36.3
Milk and other Milk Products	7.9	7.3	15.1	9.2	14.0	10.2	16.5	12.0	18.8	10.9	14.0	10.9	11.4	10.2
Tea, Sugar etc.	10.3	10.5	9.0	7.9	8.8	7.6	8.7	5.9	11.5	7.9	8.2	8.1	8.3	8.6
Other food	9.8	19.4	9.9	17.4	10.7	15.8	10.5	21.1	11.4	15.8	8.7	17.0	9.5	17.0
Total food	77.6	77.0	71.2	68.0	70.6	69.7	73.2	66.2	76.8	77.7	65.3	70.6	68.8	72.0
Clothing	10.9	10.7	10.1	14.0	10.2	10.2	10.8	11.8	9.5	9.9	11.3	11.3	10.7	12.5
Fuel and lighting	1.4	2.5	1.7	4.7	3.5	4.8	3.0	5.2	1.7	2.7	3.2	4.3	3.9	3.0
Other non-food	10.1	9.8	17.0	13.3	15.7	15.2	13.0	16.8	12.0	9.8	20.3	13.8	16.6	12.5
Total non-food	22.4	23.1	28.8	32.0	29.4	30.3	26.8	33.8	23.2	22.3	34.8	29.4	31.2	28.0
Total	100.0	100.0	100.0	100.0	100.0	100.0	100.0	100.0	100.0	100.0	100.0	100.0	100.0	100.0

Sources: Field Survey, 1965-66.
Field Survey, 1974-75.

income levels preferences shifted with increase in incomes more in favour of non-foodgrain items of food while at upper income level, shift was prominent in favour of clothing and fuel. Improved standard of living or enrichment for small income earners was more in regard to 'variety' added to food, in the case of upper income groups, enrichment manifested more in favour of clothing and lighting. Small farmers' behaviour conforms to one observed in regard to agricultural labourers.

For non-farm occupation groups, individually for groups of artisans, traders and service persons, in 1965-66 the share of expenditure on food items was high. Decline in share of food expenditure over 1965-66 and 1974-75 was experienced by all groups, the shift within food articles was sharply in favour of 'other food items' and decline in share of expenditure on foodgrains and milk and milk products was more than modest. The shifts in regard to non-food items differed for different groups. Artisans, for instance, experienced a shift more in favour of clothing and fuel, for salary earners the shift was more in favour of lighting and others, traders had modest shifts is favour of all the three items. We noted in 1965-66 most non-farm occupation groups were economically less better off than big or even medium farmers. Relative prosperity of non-farm groups is reflected therefore more in terms of increased allocation of expenses on choice foods than that on non-food items.

Consumption: Levels and Pattern (Regionwise)

We briefly discuss different aspects of level and pattern of consumption for different regions. Table 12.5 gives details of overall level of annual expenditure per capita for 1965-66, 1974-75 at current and constant prices for three regions. This table gives relative value of levels of expenditure for Matar and Limbasi, compared to the level of overall expenditure in Navagam.

All the three regions show a decline in per capita real expenditure if cost of living index for rural labour for Gujarat is used as a deflator, while all the regions would show an increase if paddy price is used. An important observation, arises from the data, we find that over 1965-66, to 1974-75, the inter-regional differences in expenditure level have significantly narrowed, even Limbasi and Matar have come closer. The pattern of consumption is indicated by budget shares. Table 12.6 gives details of budget shares for three regions for 1965-66 and 1974-75.

TABLE 12.5: Per Capita Expenditure at Current and Constant Prices - Regionwise

(*Rupees*)

	Regions			All Regions
	Navagam	Matar	Limbasi	
1	2	3	4	5
1965-66	462	337 (72.9)@	317 (68.6)@	373
1974-75 (Current)	816	723 (88.6)@	692 (84.8)@	743
PPD* (Constant Prices)	489	433	414	
COLI** (Constant Prices)	348	308	295	317

@ Figures in parentheses are relative in terms of percentage to expenditure level for Navagam for respective years.

* At constant prices obtained by using paddy price as a deflator.

** At constant prices obtained by using cost of living of agricultural labour for Rural Gujarat.

Two interesting observations emerge from Table 12.6. Engel curve behaviour applies in its strictest sense to cereals. We find firstly, that in all the three regions share of expenditure on cereals declined over the past decade. Secondly, in 1974-75 shares of cereals in total expenditure observed in Navagam and Matar were very close, that in Limbasi was much above other two regions. The other major finding pertains to share of expenditure on food to total expenditure. In Navagam it increased, in other two regions, the share remained unchanged. A decline in share of cereals was more than matched by increased share of expenditure on sugar, *gur*, tea (not necessarily nutritious items) and other foods which include besides vegetables and fruits, eggs and meat, such items as oil, spices etc. The latter items have larger share in expenditure. In all the three regions share of expenditure on milk declined, decline was marked in Navagam. Thus on the whole, the pattern of expenditure on food underwent a recognisable change in the three regions, it was accompanied by a rise in share of overall expenditure on food as in Navagam, or by no change as observed in the other two regions.

In the case of non-food items, budget shares showed an increase in what might be considered basics, like clothing and fuel and lighting. Share of expenditure on 'other' non-food items which excluded education, health, travel, entertainment, cosmetics, showed a decline. In Navagam both the increases and decreases were sharper than these observed in Matar and Limbasi.

TABLE 12.6: Budget Shares: Regions

(Per cent)

Commodity Groups	Regions							
	Navagam		Matar		Limbasi		All Regions	
	1965-66	1974-75	1965-66	1974-75	1965-66	1974 75	1965-66	1974-75
1	2	3	4	5	6	7	8	9
Cereals	36.7	34.5	40.7	34.6	43.5	41.6	40.9	36.3
Milk and Milk Products	13.7	7.28	11.7	10.6	12.2	10.6	11.7	10.2
Sugar, Tea	8.4	12.4	8.08	12.9	8.74	12.0	8.5	12.5
Other food	8.15	14.2	9.88	13.7	10.1	10.5	10.2	13.1
Total food	67.0	70.4	71.2	71.8	74.5	74.8	71.3	72.1
Clothing	9.4	12.6	10.7	12.5	10.2	12.1	11.1	12.5
Fuel and lighting	2.84	3.71	1.65	2.83	1.25	2.30	1.9	2.59
Other non-food	20.7	13.2	16.5	13.0	13.1	10.9	15.7	12.9
Total non-food	33.0	29.6	28.8	28.3	24.6	25.3	28.7	28.0
Total	100.0	100.0	100.0	100.0	100.0	100.0	100.0	100.0

Thus on the whole, more than level, pattern of consumption showed significant changes. While overall allocation between food and non-food groups was not changed in Matar and Limbasi, within-group changes were important, improved quality of food, and not so improved quality of non-food consumption. In Navagam shift was in favour of food, mainly better food, clothing and fuel. The change is thus mixed, would not enable us to draw an inference regarding overall trend.

Summary

The analysis of current consumption expenditure does provide an indication of a change for better. Depending on the deflator used we find per capita consumption expenditure to have marginally declined or marginally improved. But underneath this overall measure lies valuable observations, namely the decline, if real, is observed for big and medium farmers, and gain for all other groups including the weaker sections of labour, artisans and small farmers. For all groups together, distribution, (in terms of Gini ratio) seems to have become less unequal and the value of the inequality coefficient which was low has been further reduced. The regional analysis shows narrowing of inter-regional differences in per capita expenditure levels. If shows also a change in pattern of expenditure and pattern of expenditure distribution and poverty levels such that the regions have come closer.

On the whole there is a welcome shift in pattern of consumption, in favour of choice food and non-food items for weaker sections and in favour of non-food items for all other groups including even big and medium farmers who seem to have suffered a reduction in current consumption.

Comparing consumption expenditure measured in current prices at two points of time inevitably involves a use of price deflator. Relevance of a given price deflator would always be debatable. Our exercise shows that both for measuring a change in the level of poverty, selection of a deflator has a crucial role to play. While aggregate expenditure per capita is a meaningful indicator budget shares portray more sustained changes in the choices for the pattern of living. By this criterion we find the change for the better.

APPENDIX 12.1

Per Capita Annual Expenditure in Rs. (0.00) by Occupation Groups and by Commodity Groups (1965-66)

(All Regions)

Sl. No.	Commodity	Cultivators					All Cultivators	Non-Farm Households						
		0-5	5-10	10-15	15-25	25 and above		Agricultural Labourers	Artisans	Business	Service	Miscellaneous	Non-Farm	All
1	2	3	4	5	6	7	8	9	10	11	12	13	14	15
1.	Foodgrains	130.70	158.32	186.89	188.83	239.73	163.65	106.97	121.60	122.33	134.09	122.14	126.99	144.62
2.	Milk and Milk Products	25.49	48.46	58.87	61.16	79.53	46.03	17.03	38.12	54.51	50.89	61.55	51.82	41.44
3.	Edible Oil	11.15	16.46	24.32	22.58	36.59	18.32	9.43	13.45	18.51	18.53	14.78	16.10	14.81
4.	Sugar, tea, etc.	24.77	31.56	38.93	36.79	52.64	32.81	22.10	28.22	31.42	29.13	31.78	30.22	30.12
5.	Meat, eggs	1.42	2.58	3.81	1.37	8.17	2.77	1.00	0.00	1.27	1.30	0.54	0.98	2.04
6.	Vegetables	15.89	19.87	23.67	23.83	33.06	20.68	10.78	15.97	17.62	16.55	13.21	15.91	17.70
7.	Fuel and Light	3.9	6.50	9.65	7.92	10.15	6.57	2.96	6.97	16.73	14.77	4.95	11.84	14.21
8.	Clothing	28.29	38.91	57.79	51.41	82.60	43.50	23.52	39.96	44.22	42.56	36.22	41.67	39.12
9.	Footwear	4.78	6.60	7.55	7.99	9.20	6.48	4.06	4.66	5.46	6.02	6.76	5.83	5.86
10.	Education	2.92	6.66	19.72	22.70	31.00	11.29	0.98	5.22	16.62	18.20	4.79	14.22	9.84

Contd.

Contd.

1	2	3	4	5	6	7	8	9	10	11	12	13	14	15
11.	Medicine	5.44	20.70	17.32	20.57	8.50	13.50	2.48	10.36	9.81	11.70	6.74	9.16	10.39
12.	Entertainment	0.23	0.62	1.55	1.51	2.03	0.85	0.08	0.36	1.34	2.30	0.34	1.19	0.77
13.	Travelling	2.19	4.58	6.43	11.96	10.52	5.29	0.81	4.33	14.39	6.21	5.16	7.96	4.95
14.	Miscellaneous	0.76	4.82	4.28	7.39	14.29	4.40	0.29	4.08	8.35	10.59	8.42	6.29	3.97
15.	House Rent	0.17	1.13	0.00	2.93	1.00	0.76	1.15	5.37	12.03	13.72	1.10	8.88	2.54
16.	House repairing	1.36	6.32	12.12	9.28	0.10	5.19	0.53	5.38	2.87	9.52	3.59	4.94	4.32
17.	Charity	0.32	2.81	4.02	5.87	11.88	3.26	0.12	0.23	2.07	1.30	2.05	1.35	2.23
18.	Books and Newspapers	0.00	0.07	0.15	0.22	0.57	0.12	0.00	0.12	0.94	0.58	0.41	0.49	0.17
19.	Soap and toilet articles	2.39	3.30	4.74	5.66	6.05	3.69	1.50	2.74	5.68	5.26	2.50	4.33	3.39
20.	Tobacco	11.28	12.57	15.28	13.18	23.04	13.59	9.69	6.35	10.50	10.16	9.13	9.68	11.99
21.	Total food	209.42	277.25	336.48	334.54	449.71	284.26	167.31	217.36	245.66	250.29	244.00	241.11	250.73
22.	ONF	31.85	70.26	93.75	109.26	118.16	68.43	21.69	49.25	90.05	95.57	46.00	74.92	60.44
23.	Total non-food	64.08	115.67	160.59	168.59	210.91	118.50	48.17	96.18	151.00	153.20	87.17	128.43	113.75
24.	Total Expenditure	273.50	392.92	497.07	503.13	660.63	402.75	215.48	313.54	396.66	403.49	331.17	369.54	364.48

Source: Field Survey, 1965-66.

APPENDIX 12.2

Per Capita Expenditure in Rs. (0.00) Per Annum by Occupation Groups and by Commodity Groups in 1974-75

(All Regions)

Sl. No.	Commodity	Cultivators					Non-Farm Households							
		0-5	5-10	10-15	15-25	25 and above	Agricultural Labourers	Artisans	Traders	Servicemen	Misc.	All Cultivators	All non Farm households	All households
1		2	3	4	5	6	7	8	9	10	11	12	13	14
1.	Rice	108.46	116.31	117.20	122.59	127.63	93.65	102.24	143.66	96.81	121.31	113.37	113.11	109.47
2.	Millets	58.42	56.90	43.38	51.19	53.27	69.30	46.33	68.18	44.53	9.09	56.12	61.99	59.41
3.	Kodra	0.42	1.55	0.81	0.00	0.0	0.85	0.00	0.00	0.21	4.62	0.77	1.42	0.87
4.	Bawto	2.85	1.59	1.00	0.00	0.00	1.57	0.00	1.50	0.25	0.44	1.94	0.47	1.69
5.	Jowar	1.91	1.24	1.74	3.69	0.00	3.49	4.60	0.00	1.72	2.65	1.64	2.25	2.08
6.	Wheat	51.35	55.17	53.47	63.31	75.48	78.18	65.29	83.76	58.61	99.82	64.49	76.29	61.81
7.	Pulses	34.17	29.30	29.22	30.31	31.83	25.77	30.38	54.67	28.54	48.60	31.76	39.27	31.52
8.	Others	2.81	3.03	0.83	1.88	1.98	2.86	0.32	8.19	1.62	3.25	2.51	2.98	2.63
9.	Total cereals and pulses	260.35	265.09	252.66	272.57	295.18	275.67	249.17	359.96	232.29	369.78	262.59	297.77	269.50
10.	Milk	56.73	62.71	63.18	72.01	68.40	49.76	57.84	75.87	77.37	78.95	61.02	73.81	60.40
11.	Ghee	14.59	15.20	22.28	38.16	42.37	0.98	10.82	25.86	24.56	14.93	18.34	19.32	15.07
12.	Other Milk Products	0.14	0.0	0.0	3.13	0.0	0.15	0.00	0.00	0.00	0.00	0.19	0.00	0.16
13.	Hydrogented-oil	7.86	8.10	8.22	5.69	1.91	3.77	10.85	14.06	10.80	7.90	7.68	10.52	7.27
14.	Edible oil	33.00	34.23	34.75	38.14	35.75	37.56	40.37	44.02	39.31	35.84	34.07	39.31	35.40
15.	Gur	20.77	19.38	20.90	19.88	21.48	26.63	17.50	23.18	22.47	28.19	20.34	23.31	21.94
16.	Sugar	19.47	18.86	17.72	17.05	17.97	20.25	21.77	23.69	19.61	18.61	18.91	20.44	19.36
17.	Mutton	6.24	5.59	3.75	4.07	4.98	5.33	1.00	10.11	6.48	6.21	5.78	5.97	5.71

Contd.

Contd.

1	2	3	4	5	6	7	8	9	10	11	12	13	14
18. Eggs	0.54	1.00	0.00	0.00	0.00	1.95	0.40	0.00	1.83	0.20	0.60	0.76	0.88
19. Fresh vegetables	13.20	5.93	3.60	2.00	10.29	7.71	16.42	28.61	21.26	8.00	9.15	17.72	9.93
20. Dry vegetables	28.74	34.89	42.10	39.97	28.57	38.72	30.22	27.51	48.01	43.20	32.89	39.64	34.87
21. Fresh fruits	0.59	1.30	1.11	0.0	0.60	0.23	0.50	1.83	1.71	0.00	0.90	1.00	0.78
22. Dry fruits	1.20	1.67	1.09	4.31	0.0	0.50	0.00	0.00	0.90	0.00	1.39	0.31	1.08
23. Tea and coffee	25.60	18.52	17.96	16.96	20.02	25.78	19.87	28.45	30.24	21.28	21.74	25.31	22.97
24. Salt	4.79	0.00	0.00	0.00	0.02	0.04	0.00	0.17	0.03	0.21	2.22	0.10	1.53
25. Condiments and spices	25.53	22.26	22.72	20.13	21.69	37.79	30.21	29.58	26.27	33.17	23.87	29.62	27.30
26. Other food	1.32	0.74	0.76	0.00	8.59	0.51	0.00	1.44	0.08	0.68	1.38	0.48	1.10
27. Total food	520.92	517.20	512.80	554.46	577.84	533.42	506.97	694.36	563.20	667.15	523.40	605.39	535.46
28. Clothing	88.56	95.69	102.59	110.48	158.37	74.13	104.34	101.46	100.03	84.61	97.29	96.60	92.67
29. Fuel and light	15.89	17.04	20.48	17.19	34.07	16.78	29.00	40.61	35.04	20.51	17.93	30.59	19.28
30. Electricity	2.43	3.11	4.60	4.37	4.04	0.71	5.90	7.61	9.35	2.95	3.11	6.51	3.07
31. Health	1.92	16.54	14.13	18.59	36.65	9.37	8.65	28.32	18.87	13.27	15.39	16.89	14.40
32. Education	7.51	7.05	14.29	28.34	15.41	1.76	13.38	11.62	13.49	6.06	9.50	10.98	8.17
33. Reading materials	0.59	1.26	2.55	0.33	0.85	0.10	0.85	1.36	3.34	0.78	1.05	1.77	0.96
34. Entertainment	6.16	5.89	9.76	15.75	12.01	4.49	5.92	6.36	8.70	4.53	7.16	6.54	6.56
35. Communication	2.10	2.62	3.11	5.33	5.10	1.26	2.71	4.94	3.68	3.04	2.69	3.55	2.52
36. Toilet articles	21.21	23.65	26.12	37.45	32.06	18.92	27.56	32.54	31.95	23.50	23.79	28.73	23.45
37. Tobacco	19.14	19.46	23.32	25.69	35.84	23.65	20.71	27.07	14.31	25.21	20.97	20.93	21.49
38. Travel	1.01	2.14	2.61	1.44	16.27	0.39	1.24	2.78	0.87	0.33	2.47	1.11	1.90
39. House Rent	0.46	0.12	0.74	0.00	0.00	1.50	11.09	24.84	32.01	1.15	0.34	17.72	2.72
40. Recreation and Social Expenditure	11.81	7.34	7.34	18.75	16.03	5.52	6.32	10.14	12.98	4.56	10.31	8.74	9.18
41. Others	1.68	1.27	2.56	0.00	3.21	1.22	0.95	1.72	2.79	1.36	1.65	1.83	1.59
42. Total non-food expenditure	190.47	203.18	234.14	283.76	369.90	159.80	238.63	301.35	287.41	191.96	213.68	252.50	207.95
43. Total expenditure	711.39	720.38	746.93	838.21	947.74	693.21	745.55	995.71	850.60	859.11	737.08	857.89	743.37

Source: Field Survey, 1974-75.

13

WEALTH AND WELL-BEING

WE HAVE examined levels of consumption in money and real terms and impact of irrigation on them, mainly through expenditure on articles of current consumption. Expenditure on items of current consumption is not the whole, but a part of the total consumption expenditure. Account of expenditure will not be complete unless one also considers expenditure on articles which are not used up when they are consumed, these articles are consumer durables. Their 'consumption' is in terms of use of their services or to put it in economic terminology, we consume a flow of their services. Our period of reference for accounting is one year. Some articles like earthenwares or clothes would be used up or 'consumed' during the year. We have taken expenditure on clothing into account while discussing current consumption levels but not expenditure on earthenwares. The line dividing current consumption from consumption of durable articles is thin and to an extent conventional.

The expenditure on durable consumption goods will depend on their durability, less durable a good, greater the frequency of expenditure on them for their replacement. Within a period of one year we may therefore find expenditure to be incurred on less durable goods by relatively larger percentage of households and *vice versa*. If we go by expenditure incurred on items during a year we would get relatively very few observations which may distort comparisons across families and over time. More often than not, such expenditure on durable assets will represent the transitional component. When such component is positive and large in a current year's income compared to more permanent income expenditure on durable goods will be large. Such transitional component may vary for different families and good proxies for transitional components or permanent incomes may be difficult to come by for individual families. Land area, ignoring its quality, is admittedly a poor proxy for permanent income at individual household level. Besides, for non-cultivating households, this proxy will not be of avail.

We should note however another major lacuna in expenditure analysis. Expenditure may be incurred for social events and festivities. In rural areas (as also in urban areas) weddings claim a fair share in total expenditure of the families celebrating weddings. But the same problem of randomness and small number of observations will limit usefulness of detailed analysis regarding wedding expenses. Besides, since the expenditure does not result into an acquisition of a physical asset, (except ornaments and jewellery information about which is difficult to obtain), the status of any household would change from one year to the next, since marriages will not be performed year after year in a family. Hence, we decided to exclude social expenditure from the analysis of well-being. It appears however, to a limited extent, as one of the purposes of borrowing while analysing the burden of debt.

Considering these difficulties we decided to measure not expenditure, not even quantity, but relative status of households on selected items of durable goods mainly taking 'yes' or 'no' position, namely whether given family had or did not have a specified item in its possession and in use during the year. Relative positions of households varied for 1965-66 and 1974-75. We give lists of durables for the respective years in the appendix, we also give a detailed description of methodology adopted for the present analysis. The main features of the methodology are indicated below:

1. We take the fact of possession and not quantity of each item in our account.
2. We omit those items which are owned by all households, since all households will have equal relative status.
3. Items not owned at all or owned by very few viz. less than 5 per cent, are omitted since most, if not all, families would have equal relative status of not owning that item.

Based on methodology described in the appendix we evolved an index. This is strictly a relative and not an absolute measure of 'well-being' of the household, it is relative to the group of all sample families taken together. When we compared for any subgroup its relative positions at two points of time, we would be able to measure relative distance of the subgroup from the other subgroups and observe whether it is increasing or narrowing. Its main limitation is, we are unable to measure absolute improvement of well-being over time of the entire group of sample families or for individual subgroups, we fill up partly this gap through item analysis. Simple tables giving comparisons of percentage of families owning individual items at two points of time are presented in Table 13.2. We begin with them.

Consumer Durables : Item Analysis

We listed 48 items of consumer durables in 1965-66 and 81 items in 1974-75. We have thus expanded the list in 1974-75. We have a limited range about 32 items, which are common for 1965-66 and 1974-75. Metal drums used for storage have been dropped from analysis. 32 items in 1965-66 however would be equivalent to more than 32 items in 1974-75, since such broad categories as wooden furniture have been expanded to include specified individual items. Such broad categories are (1) Wooden furniture (includes cane furniture); (2) Kitchen wares and (3) Electrical appliances and others. The individual items in 1974-75 have been collapsed into suitable broad categories for comparisons. There were 28 to 30 new items included in the list of consumer durables for 1974-75, they were not canvassed for 1965-66.

With the above limitations, we give, to begin with, details of percentage of households possessing specific durable assets in 1965-66 and in 1974-75. An unweighted average at two points of time suggests that spread of use of different assets has increased by 48 per cent in 9 years' time which is a substantial improvement. A few typical instances may be cited. For instance in rural areas wooden bed (all wood) is not commonly used. It is a status good.

In 1965-66 only 5 per cent of households had it, in 1974-75, its possession spread to 35 per cent of households. Use of glassware (tumbler etc.) was rare in 1965-66, confined to 18 per cent. It now obtains for 68.5 per cent of households (Table 13.2). Use of copper and brass vessels is now nearly universal and that of aluminium vessel, an 'inferior good' declined, from 99.00 to 84.07 per cent, i.e., about 15 per cent of households gave up the use of aluminium vessel. There are several items which would give similar trends, we can mention instances of suit-cases, fountain pens/ball pens, wrist watches, radio, battery torches, use of electricity, lantern, even steel furniture, and wooden furniture. Surprisingly the percentage of households using alarm clock, wall clock, cycle, sewing machine, has not shown a marked increase.

TABLE 13.1: Percentage* of Families Using Consumer Durables by Occupation (1965-66 and 1974-75)

Occupation group	Percentage of families using consumer durables in			
	1965-66	1974-75	(3)–(2)	Percentage change 4/2 × 100
1	2	3	4	5
Cultivators (acres)				
0 to 5	13.2	36.6	23.4	176.9
5 to 10	21.9	43.1	21.2	97.1
10 to 15	35.3	52.4	17.1	48.3
15 to 25	36.8	54.0	17.2	46.7
25 +	43.5	53.6	10.1	23.2
All Cultivators	22.9	43.4	20.6	89.9
Agricultural Labourers	15.8	27.2	11.4	72.4
Artisans	30.5	45.6	15.1	49.4
Traders	39.4	49.9	10.5	26.7
Services	32.6	51.0	18.4	56.5
Others	22.1	35.7	13.5	61.2
All non-farm households	30.5	45.6	15.1	49.5
All Households	27.2	40.3	13.2	48.4

* These percentages are simple arithmetic mean of the percentage given for each consumer durable item groups. Values thus obtained by themselves speak neither for the exact coverage nor for the quality. They only show whether the mean value of use has gone-up in 1974-75 or not. Secondly, they enable, though crudely, a rough comparison among the cross-sections at two points of time.

Sources: Field Survey, 1965-66.
Field Survey, 1974-75.

The new items like teapoy, camera, sofa, telephone, dressing table, plasticwares have made an entry in rural life, though their use is restricted to a few households. There are certain items like cradle, musical instruments, which were omitted in 1965-66, but were canvassed in 1974-75. Of the two, cradle seems to be owned by over 58 per cent of houses, the traditional method of cradling a child would be an improvised cloth piece tied to pillars, or tree trunks or branches of two trees.

TABLE 13.2: Percentage of Households Owning Consumer Durables to Total Households — Matar Taluka

Sl. No.	Consumer durable	Percentage of households using consumer durables in		Percentage change
		1965-66	1974-75	(3)–(2) ÷ (2) × 100
1		2	3	4
1.	Brass vessels	98.4	100.0	+ 1.6
2.	Copper vessels	79.5	93.8	+ 14.3
3.	Aluminium	99.2	84.1	− 15.1
4.	Steel	9.8	39.6	+ 29.8
5.	Glasswares	28.4	68.5	+ 40.1
6.	Chinaware	62.8	84.0	+ 21.2
7.	Wooden Bed	5.0	35.2	+ 30.2
8.	Newar Bed	27.7	38.8	+ 11.1
9.	Coir Bed	89.0	86.0	− 3.2
10.	Steel Bed	3.5	12.0	+ 8.5
11.	Wooden furniture	16.0	30.2	+ 14.2
12.	Steel furniture	2.4	10.3	+ 7.9
13.	Cane furniture	0.6	—	—
14.	Lantern	59.3	79.8	+ 20.5
15.	Petromax	5.4	5.0	− 0.5
16.	Electric light	10.1	27.4	+ 17.3
17.	Torch	21.4	49.1	+ 27.7
18.	Coal stove	9.9	14.0	+ 4.1
19.	Kerosene stove	25.0	46.2	+ 21.2
20.	Wrist watch	15.6	39.6	+ 24.0
21.	Alarm clock	8.6	8.6	+ 0.0
22.	Wall clock	3.4	10.0	+ 6.6
23.	Radio	7.2	17.7	+ 10.5
24.	Sewing machine	1.2	1.4	+ 0.2
25.	Thermos	2.1	5.0	+ 2.9
26	Cycle	11.6	13.3	+ 1.7
27.	Motor Cycle/Car	0.3	1.5	+ 1.2
28.	Fountain pen/Ball pen	33.8	57.0	+ 23.2
29.	Umbrella	68.9	83.8	+ 14.9
30.	Rain coat	4.6	3.9	− 0.7
31.	Suit Case	6.3	38.1	+ 31.8
32.	Electric fan	2.6	6.9	+ 4.3

Sources: Field Survey, 1965-66.
Field Survey, 1974-75.

Relative Positions of Occupation Groups

While for all groups, for listed items-owners were 27 per cent in 1965-66, and 40 per cent in 1974-75. For agricultural labourers, economically the weakest, the percentage of households owning the listed items increased from 16 in 1965-66 to 27 in 1974-75 (Table

13.3). In the case of labourers sharp increase in regard to consumer durables compares well with the improvement in current consumption expenditure in real terms.

We find similar improvement for small farmers. Percentage of small farmers owning consumer durables increased from 13.2 in 1965-66 to 36.6; for medium and big farm households we find a relieving situation while their real current consumption expenditure seems to have declined in 1974-75 compared to that in 1965-66 (Table 13.4) their consumer durables status shows an improvement over already fairly high initial position. In 1965-66 for listed items, percentage of big farmers was 43, it increased to 54 in 1974-75, similarly for two groups, 10-15 acres and 15-25 acres size classes, the percentage improved from 35 and 37 respectively in 1965-66 to 52 and 54 respectively. For all non-farm occupation groups the improvements were from 39.4 to 49.9 per cent (traders) 32.6 to 53.0 per cent (services) 22.1 to 35.7 per cent for miscellaneous occupations group. In general introduction of irrigation seems to have improved wealth and well-being of all classes, including the weakest, improvement being relatively greater for weaker classes. Further, those whose position seems to have worsened regarding real current consumption expenditure, also show an improvement implying that if seasonal effects are excluded irrigation expansion seems to have given benefit to all classes and also seems to have narrowed inequality in ownership of consumer wealth. A few instances may be given to illustrate the nature of change. For instance no labour family possessed any stainless steel utensil in 1965-66, while in 1974-75, 13 per cent of them owned one or more of these utensils. Among small farmers the percentage of stainless steel utensils owners went up from 2 to 36. For wooden beds (cots) the percentage of owners increased among labourers from 1 to 35 and among small farmers from 2 to 32. Electric lighting obtained for 4 per cent of small farmers in 1965-66 but for 25 per cent in 1974-75. Labourers are behind small farmers, only 5 per cent of them had electricity connection. Labourers did not possess in 1974-75 an alarm clock, a sewing machine, a thermos flask. Radio is quite common with the advent of transistors, 50 per cent of big farmers owned them in 1974-75 and 25 to 35 per cent of different groups of medium farmers (5 to 25 acres) owned them. Among small farmers percentage of owners was 12, but among labourers, it was less than three. Though there is an improvement, labourers are still far behind the cultivating classes. On the whole relative positions of different

TABLE 13.3: Percentage of Households Owning Consumer Durables in 1965-66 and 1974-75

Sl. No.	Item	Occupation groups — Cultivator households (Size of holding in acres)									
		0 to 5		5 to 10		10 to 15		15 to 25		25 and above	
		1965-66	1974-75	1965-66	1974-75	1965-66	1974-75	1965-66	1974-75	1965-66	1974-75
1	2	3	4	5	6	7	8	9	10	11	12
1.	Brass utensils	100.0	99.0	100.0	98.3	100.0	95.9	100.0	100.0	100.0	100.0
2.	Copper utensils	89.9	97.4	92.9	97.2	96.2	93.2	93.0	100.0	97.1	100.0
3.	Enamel Utensils	100.0	86.6	99.4	87.2	98.7	82.2	100.0	95.7	100.0	71.4
4.	Steel utensils	2.2	35.5	7.0	46.7	24.1	56.2	28.0	78.3	35.2	57.1
5.	Glassware	19.5	59.0	31.9	69.4	45.7	68.5	48.9	91.3	61.89	96.4
6.	Chinaclay ware	61.8	85.3	62.3	88.9	81.0	76.7	79.1	87.0	83.3	82.1
7.	Wooden Bed (cots)	1.7	31.9	1.8	36.7	7.7	43.8	9.3	43.4	17.6	53.6
8.	Newar cots	21.3	33.2	30.3	47.3	46.8	52.0	48.8	73.9	61.8	67.9
9.	Coir cots	96.9	90.6	95.5	85.0	87.3	84.9	90.1	91.3	82.3	89.3
10.	Steel cots	0.0	8.5	1.3	20.6	7.6	20.6	14.0	26.1	17.6	25.0
11.	Wooden furniture	7.0	25.4	14.2	35.0	44.5	46.6	32.6	65.2	52.9	39.3
12.	Steel furniture	0.0	6.5	1.8	1.7	3.9	21.9	11.7	30.4	8.8	32.14
13.	Cane furniture	0.4	0.0	0.0	0.0	1.3	0.0	0.0	0.0	2.9	0.0
14.	Huricane lantern	53.6	83.1	71.8	89.4	81.0	100.0	83.7	86.0	85.3	71.4
15.	Petromax	1.8	3.6	3.2	5.0	14.0	6.9	18.6	4.4	32.4	42.9
16.	Electric lighting	3.5	25.4	10.4	30.6	12.8	42.5	21.0	47.8	14.6	35.7
17.	Torch	12.0	51.1	27.8	70.6	38.2	67.1	53.5	82.6	73.5	82.1

Contd.

TABLE 13.3 *Contd.*

1	2	3	4	5	6	7	8	9	10	11	12
18.	Coal stove	3.0	10.1	12.3	20.6	19.0	21.9	16.3	17.4	14.6	10.7
19.	Primus	9.3	42.4	25.2	48.3	45.7	60.3	39.6	65.9	61.7	64.3
20.	Wrist watch	4.3	32.9	15.5	56.1	35.5	56.2	32.6	60.9	49.0	71.4
21.	Alarm clock	1.3	5.2	9.0	15.0	21.5	13.7	25.5	13.0	38.3	17.9
22.	Wall clock	1.6	6.5	1.2	15.0	7.6	20.6	9.3	30.4	11.7	32.1
23.	Radio	1.2	11.7	7.1	22.2	7.6	35.6	23.3	34.8	29.5	50.0
24.	Sewing machine	0.0	0.7	0.6	1.7	1.3	2.7	2.3	0.0	0.0	0.0
25.	Thermos flash	0.4	3.3	1.3	8.9	6.3	11.0	2.3	17.4	17.7	32.1
26.	Cycle	3.5	11.1	14.8	21.1	27.9	27.4	23.2	26.1	38.3	28.6
27.	Motorcycle/Car	0.0	0.7	0.0	2.2	0.0	5.5	2.3	4.4	0.0	7.1
28.	Fountain pen/Ball pen	20.7	49.5	38.3	66.1	52.1	69.9	55.9	73.9	67.6	82.1
29.	Umbrella	74.7	82.1	85.7	87.8	94.9	90.4	95.3	91.3	88.2	96.4
30.	Raincoat	1.3	1.0	1.9	4.4	7.7	5.5	2.3	8.7	14.6	21.4
31.	Suitcase/Leather bag	0.8	31.6	3.2	38.9	12.6	48.0	6.9	56.5	23.5	78.6
32.	Electric fan	0.8	2.0	1.3	8.3	6.3	17.8	7.0	21.7	8.8	21.4
	All durables (simple average)	13.2	36.6	21.9	43.1	35.3	52.4	36.8	54.0	43.5	53.6

TABLE 13.3 (contd.)

Sl. No.	Item	Occupation Groups — Agricultural labourers and Non-Cultivator Households											
		Agricultural labours		Artisans		Traders		Service		Others		All households	
		1965-66	1974-75	1965-66	1974-75	1965-66	1974-75	1965-66	1974-75	1965-66	1974-75	1965-66	1974-75
1	2	13	14	15	16	17	18	19	20	21	22	23	24
1.	Brass utensils	95.0	99.6	100.0	100.0	100.0	97.3	100.0	100.0	95.2	100.0	98.4	100.0
2.	Copper utensils	52.5	81.9	91.2	91.7	87.8	89.5	63.1	97.3	82.4	92.7	79.5	93.8
3.	Enamel utensils	99.2	76.0	97.1	83.3	97.6	86.8	97.5	87.8	100.0	89.9	99.2	84.1
4.	Steel utensils	0.0	12.6	5.8	58.3	31.7	57.9	22.2	77.0	8.0	29.4	9.8	39.6
5.	Glassware	9.8	62.0	44.1	80.6	51.4	76.3	49.2	91.9	20.7	64.7	28.4	68.5
6.	Chinaclayware	48.4	75.8	73.5	86.1	87.8	81.6	65.4	91.9	54.0	77.9	62.8	84.0
7.	Woodenbeds	1.2	34.7	11.7	25.0	7.3	47.4	17.2	18.9	4.8	32.4	5.0	35.2
8.	Nevar cots	6.5	17.3	23.5	50.0	43.8	57.9	44.4	58.1	20.7	35.3	27.7	38.8
9.	Coir cots	92.2	90.0	79.4	29.4	78.0	79.0	59.3	60.8	87.2	82.4	89.0	85.8
10.	Steel cots	0.4	1.5	5.8	11.1	7.3	13.2	4.9	18.9	0.0	8.8	3.5	12.0
11.	Wooden furniture	2.4	11.8	23.5	44.4	36.7	63.2	24.6	54.1	12.8	22.1	16.0	30.2
12.	Steel furniture	0.0	3.3	2.9	11.1	7.3	10.5	6.1	32.4	1.6	7.4	2.4	10.3
13.	Cane furniture	0.0	0.0	0.0	0.0	2.4	0.0	1.2	0.0	0.0	0.0	0.6	0.0
14.	Huricane lantern	37.3	61.2	73.5	86.2	92.7	86.8	56.8	67.6	47.8	76.5	69.3	79.8
15.	Petromax	0.0	1.1	5.8	2.8	17.1	15.8	5.0	0.0	3.2	1.5	5.4	5.0
16.	Electric lighting	0.8	5.5	20.6	41.7	36.6	50.0	29.6	68.9	8.0	17.7	10.1	27.4
17.	Torch	2.0	19.2	23.5	44.4	24.4	79.0	34.5	52.7	11.2	39.7	21.4	49.1

Contd.

TABLE 13.3 *Concld.*

1	2	13	14	15	16	17	18	19	20	21	22	23	24
18.	Coal Stove	0.4	1.5	14.7	38.9	34.5	23.7	25.8	25.7	9.6	15.8	9.9	14.0
19.	Primus	4.5	23.0	60.8	81.6	70.6	86.8	51.6	82.4	20.8	36.8	25.0	46.2
20.	Wrist watch	0.4	12.2	17.5	55.6	44.0	71.1	41.9	79.7	8.0	27.9	15.6	39.6
21.	Alarm clock	0.0	0.0	2.9	13.9	24.0	15.8	25.9	20.62	6.8	8.8	8.6	8.6
22.	Wall clock	0.0	0.0	0.0	2.8	14.6	23.7	2.5	20.3	1.6	4.4	3.4	10.0
23.	Radio	0.0	2.6	8.8	27.8	22.0	34.2	14.7	46.0	6.4	8.8	7.2	17.7
24.	Sewing machine	0.0	0.0	17.6	16.7	0.0	0.0	0.0	4.1	1.6	0.0	1.2	1.4
25.	Thermos flash	0.0	0.0	2.9	0.0	0.0	2.6	6.2	8.1	0.0	1.5	2.1	5.0
26.	Cycle	1.2	1.9	5.8	8.3	36.9	31.6	19.6	23.0	4.6	5.9	11.6	13.3
27.	Motorcycle/Car	0.0	0.0	0.0	0.0	2.3	0.0	1.2	1.44	0.0	2.9	0.3	1.5
28.	Fountain pen/Ball pen	8.5	35.4	67.6	75.0	78.0	81.6	65.5	87.8	25.6	50.0	33.8	56.0
29.	Umbrella	41.4	77.9	85.3	94.4	80.4	92.1	63.1	87.8	46.1	73.5	68.9	83.8
30.	Rain coat	0.0	1.9	2.9	2.8	12.1	5.3	20.9	12.2	3.4	2.9	4.6	3.9
31.	Suitcase/leather bag	0.4	24.7	2.9	38.9	17.0	65.8	29.5	77.0	3.2	29.4	6.3	38.1
32.	Electric fan	0.4	0.0	5.8	0.0	14.6	21.1	9.9	24.3	8.0	1.5	2.6	6.9
	All durables (simple average)	15.9	27.2	30.5	45.6	39.4	49.9	32.6	51.0	22.1	35.7	27.2	40.3

Sources: Field Survey, 1965-66.
Field Survey, 1974-75.

TABLE 13.4: Index of Consumer Durables 1965-1975

Occupation groups	Regions						All Regions	
	Navagam		Matar		Limbasi		1965	1975
	1965	1975	1965	1975	1965	1975		
Size of land (acres)								
0-5	6.31	10.50	4.25	6.14	3.57	6.44	4.57	7.43
5-10	10.27	14.53	6.08	9.37	6.15	7.85	7.57	10.62
10-15	14.83	14.52	12.99	13.72	9.82	8.74	13.02	12.73
15-25	16.63	22.20	15.35	16.99	6.50	8.40	13.39	14.03
25 +	25.17	23.98	15.84	16.71	11.99	11.95	16.82	14.87
All cultivators	11.50	13.10	6.92	8.44	6.34	7.91	8.11	9.60
Agricultural Labourers	2.95	5.68	2.31	3.92	2.49	5.47	2.55	4.81
Artisan	13.36	14.66	6.29	9.42	8.45	8.77	9.21	11.31
Trader	17.66	15.35	12.06	13.87	13.09	9.19	14.66	13.14
Salary earners	15.24	15.06	10.73	14.91	14.71	10.05	13.46	14.60
Miscellaneous	6.46	7.99	5.78	5.83	5.53	7.77	5.93	7.20
All non-farm households	13.44	13.77	8.75	11.14	9.57	8.60	10.87	11.51
All households	9.99	11.25	6.04	7.69	6.14	7.56	7.35	8.81

Sources: Field Survey, 1965-66.
Field Survey, 1974-75.

groups have not changed drastically. Correlation ranks of occupational classes for 1965-66 and 1974-75 on the basis of overall ownership of listed items was 0.87.

Index of Well-being

Wealth denotes aggregate position of value of possession of assets. Well-being can be defined as a relative status. If all households possess a good, all have equal status, if as low as 5 per cent of families possess a good e.g. a car or a refrigerator, others would not feel the loss of their status, since a wide majority have equal status. Avoiding these extremes and taking account of relative positions of individual families, duly weighted, we get an index of aggregate status of well-being of the family. We should emphasise that such an index is constructed for individual households.

On the whole of index well-being has marginally improved over the decade under study. Sharp improvement would not be expected, in fact index value was lower in 1974-75, for big and medium farmers and traders. But the relative position of different occupation groups has not materially changed, rank correlation worked out to 0.95. What is important to note is that well-being index was closely related to level of current consumption expenditure in 1965-66, rank correlation being 0.85, in 1974-75 the two were less closely related, rank correlation being much lower 0.65. 1974-75 was a relatively worse year because of bad weather and big and medium farmers' production and consumption were adversely affected, but well-being index would depend on lasting economic conditions or permanent income, the two ranks would therefore differ.

Summing-up

The impact of irrigation on standard of living of different occupation classes seem to be better reflected in terms of possession of consumer durables by different households. On the whole, there was a rapid expansion in the number of households owning common assets and quite a few new assets have made their entry in the life of rural people. The improvement is across the board, all groups enjoying better position in 1974-75 compared to that in 1965-66. What seems to be a persisting feature is that the relative positions of different groups have not changed, as it has in the case of current consumption. Even though the labourers, economically and even socially the weakest class, has improved their possession of durables, they are still for behind others including small farmers. Relative status of occupation groups in terms of their well-being in 1965-66 and 1974-75 seems to be closely related for all occupation groups. There is however one change. In 1965-66 well-being status was closely related to the per capita expenditure. This relationship was less close in 1974-75 than that obtained in 1965-66. While current consumption expenditure would reflect the weather effect to a greater extent for farmers and more so for big and medium farmers, these effects would be minimal when consumer durables are considered. From the angle of consumer durables then the effect of irrigation, to repeat, is across all classes but the benefits have not changed the relative status of different groups. We should add that analysis does not take account of either quantity or value of different items owned by different occupation groups.

APPENDIX 13.1

List of Consumer Durables

1965-66

I. *Utensils*
1. Brass
2. Copper
3. Enamel
4. Steel
5. Glassware
6. Chinaclayware
7. Tin Drums

II. *Cots*
1. Wooden
2. Newar
3. Coir
4. Steel

III. *Wooden Furniture*
1. Chairs
2. Table
3. Cup-board
4. Swing
5. Others

IV. *Steel Furniture*
1. Chairs
2. Table
3. Cup-board
4. Others

V. *Cane Furniture*
1. Chairs
2. Table
3. Others

VI. *Other Articles*
1. Huricane Lamp
2. Petromax
3. Electric light
4. Torch
5. Coal stove
6. Primus
7. Gas stove
8. Coal iron
9. Electric iron
10. Wrist watch
11. Alarm clock
12. Radio
13. Gramophone
14. Sewing Machine
15. Thermos
16. Water bag
17. Hot water bag
18. Thermometer
19. Cycle
20. Motor cycle
21. Motor
22. Fountain pen
23. Umbrella
24. Leather bag
25. Others

1974-75

I. *Utensils*
1. Brass
2. Copper
3. Enamel
4. Steel
5. Glassware
6. China clayware
7. Tin drums

II. *Cots*
1. Steel
2. Wooden
3. Newar
4. Coir

III. *Wooden Furniture*
1. Chairs
2. Writing Table
3. Steel
4. Cupboard
5. Swing
6. Others

IV. *Steel Furniture*
1. Chairs
2. Tables/stools
3. Others

V. *Lighting Article*
1. Hurricane Lamp
2. Petromax
3. Torch
4. Others

VI. *Coal Iron*

VII. *Electrical Articles*
1. Iron
2. Stove
3. Bulb
4. Tubelight
5. Table fan
6. Ceiling fan
7. Bell
8. Shaver
9. Mixer
10. Water heater
11. Geyser
12. Flour mill
13. Electric amber
14. Refrigerator
15. Air-conditioner
16. Others

VIII. *Other Articles*
1. Wrist watch
2. Alarm clock
3. Wall clock
4. Radio
5. Gramophone
6. Record player
7. Sewing machine
8. Thermos
9. Water bag
10. Hot water bag
11. Leather bag
12. Thermometer
13. Cycle
14. Motorcycle/scooter
15. Motor Car
16. Truck

17. Tempo
18. Fountain and ball point pen
19. Umbrella
20. Rain coat
21. Sofa set
22 TV set
23. Teapoy
24. Camera
25. Cradle
26. Telephone
27. Dressing table
28. Mechanical churner
29. Buckets
30. Tubs
31. Jugs
32. Others
33. Musical instruments
34. Others

APPENDIX 13.2

Consumer Durable Index

There were 84 items of consumer durables. For each item information on duration of possession is also available. Based on duration, following scheme of scores has been evolved:

S = 1 if a good is owned since a year
2 if a good is owned for 1-2 years
3 if a good is owned for 2-5 years
4 if a good is owned for 5-10 years
5 if a good is owned for 10-15 years
6 if a good is owned for 15-25 years
7 if a good is owned for more than 25 years.

The above raw scores are then converted into

$$\text{Z - Scores (Z)} = \frac{S-s}{Sd(s)}$$

where
S = Score
s = mean score
sd(s) = Standard deviation of score.

Next we obtained the weighted average of the Z-Scores denoted by A.

$$A = \frac{W_j Z_j}{W_j}, \quad = W_j = \frac{N-N_j}{N}$$

where N = total number of households
N_j = total number of households possessing item j.

An index based on values of A is derived thus:

$$I = \frac{A}{M} \times 100$$

where M = Maximum score attainable by a household. M depends on number of durable goods and the largest possible duration for which each one of them could be owned (or possessed).

APPENDIX 13.3

Consumer Durable Index, Ranks Correlation

Occupation group	Rank	
	1965-66	1974-75
0 to 5	9	8
5 to 10	7	7
10 to 15	5	5
15 to 25	4	3
25 +	1	1
Agricultural labourers	10	10
Artisan	6	6
Business	2	4
Service	3	2
Others	8	9

Rank corelation = 0.95

Ranking of Consumer Durables Index and Per Capita Expenditure

Occupation group	Rank based on per capita expenditure	Rank based on consumer durable index
	1965-66	
0 to 5	9	9
5 to 10	6	7
10 to 15	3	5
15 to 25	2	4
25 +	1	1
Agricultural labourers	10	10
Artisan	8	6
Business	5	2
Service	4	3
Others	7	8

Rank corelation = 0.85

Occupation group	Rank based on per capita expenditure	Rank based on consumer durable index
	1974-75	
0 to 5	9	8
5 to 10	8	7
10 to 15	6	5
15 to 25	5	3
25 +	2	1
Agricultural labourers	10	10
Artisan	7	6
Business	1	4
Service	4	2
Others	3	9

Rank corelation = 0.65

14

OVERVIEW

Introduction

DURING 1965-66, a second survey (the first resurvey) was carried out, covering a sample of 998 families. The main objective of the first resurvey was to trace changes in the rural area of the taluka regarding production level and technique and levels of living. The resurvey reported that irrigation expansion and the use of improved seeds and fertilisers together enabled to raise crop production level to nearly twice that obtaining 35 years ago. But there was hardly any technological change. The technique of production (as implied by relative importance of factors of production and their forms) had undergone some recognisable change; mechanical/electrical devices to lift water had almost totally replaced the use of bullocks for water-lifting.

The second resurvey was intended to provide a profile of rural economy of Matar for 1974-75 so that by comparing it with one obtaining a decade ago one can observe the sustained change, if any. The survey was expected to uncover the process of the change during the intervening nine years. The main feature of the exercise is the collection of longitudinal data for as many as possible of 998 households studied in 1965-66. Of them 801 households could be contacted. The irrigated area increased from 26 per cent of gross cropped area in 1965-66 to 45 per cent of gross cropped area in 1974-75. Also seed-fertiliser revolution popularly known as green revolution also made its impact on the economy of the taluka.

The growth of dairying in the wake of the organisation of Amul Dairy in Kheda district of which Matar taluka is a part, was yet another factor that helped promote the growth of the economy of the taluka. In Matar, the number of primary milk co-operative societies and their membership increased sharply from 32 and 5,722 respectively in 1965-66 to 62 and 11,371 respectively in 1974-75. Milk collection increased from 13.3 thousand litres per day in 1965-66 to 18.3 thousand litres per day in 1974-75. Milk yield per buffalo increased from 538 litres per year in 1965-66 to 639 litres per year in 1974-75.

Thus the two exogenous developments, irrigation and milk marketing in a rare combination provided the base for transformation of the rural economy of Matar. As we shall see the development helped improving equity of income and asset distribution.

Growth : Pace and Pattern

We begin with growth of agriculture. Table 14.1 gives changes in major indicators of agricultural growth, such as irrigation, cropping intensity, cropped area and area under selected high valued crops. The data relate to two years 1965-66 and 1974-75. A change over the period is observed by comparing the data for these two years. The change would not reflect the long-term trend.

TABLE 14.1: Selected Indicators of Agricultural Growth in Matar Taluka

		1965-66	1974-75	Change in per cent
1.	Irrigation area (in '00 ha)	121	229	89.2
1a.	Irrigated area as percentage of net cropped area*	26.0	45.0	—
1.b	Irrigated area as percentage of gross cropped area	25.0	39.0	--
2.	Area sown more than once (in '00 ha)	37	63	70.3
2a.	Cropping intensity	1.08	1.15	6.5
3	Area under paddy (in '00 ha)	130	132	1.5
3a.	Prcentage to gross cropped area	26.5	27.0	—
4.	Area under wheat (in '00 ha)	69	92	33.3
4a.	Percentage to gross cropped area	14.0	18.8	—
5.	Improved varieties and HYV			
5a.	Area under improved variety (in '00 ha)	107	—	—
5.b	Area under HYV (in '00 ha)	—	193	—
5.c	Percentage (a) & (b) to gross cropped area	21.7	39.4	—

* The data on irrigated are a differ for different sources. Aggregation of data drawn from records of individual villages give the following results: Percentage of area under irrigation to net cropped area was 30.5 for 1965-66 and 52.4 for 1974-75. Data based on sample households also differ (Table 15.2b).

Source: The GIAP - Data Bank.

Irrigated area increased during the period under study by 89 per cent, its coverage of cropped area increased from 26 per cent in 1965-66 to 45 per cent in 1974-75. Area cropped more than once increased by 70 per cent and area under wheat increased by 33 per cent. Paddy and wheat gained in importance as major crops.

Pace of Growth

The three geographical regions – (1) Navagam, (2) Matar and (3) Limbasi – of the taluka represent different soil conditions and varying irrigation history. Navagam has a history of use of canal irrigation for paddy, stretching over nearly a century; the village level data show that irrigation as a percentage of cropped area has increased in Navagam from 44 in 1965-66 to 56 in 1974-75. The other two regions had lower levels of irrigation, the irrigated cropped area being 26 per cent in Matar region and 25 per cent in Limbasi region in 1965-66. But in both the regions, irrigated area increased rapidly and covered over half of the cropped area during the decade under study.

Crop yields show a bright record, despite 1974-75 being a year of low rainfall, Matar and Limbasi regions showed a rapid increase in crop yields between 1965-66 and 1974-75. In Navagam region where the increase in irrigated area was relatively smaller increase in crop yields was only marginal.

The net sown area in the taluka as well as for the sample families declined. The cropping intensity for sample households increased from 1.05 to 1.33. Consequently gross cropped area increased from 4,943 to 5,381 acres. Applying value weights for 1965-66, we find that production index for four crops increased to 195 in 1974-75. Applying value weights for 1974-75 we get crop production index of 242. A geometric mean (an ideal index – Fisher's index) would be 217. Accordingly, annual compound rate of growth of crop production would be 7.7, 10.3 and 9.0 per cent per year. After making an allowance for decline in the share of crops not accounted for, from 16 per cent in 1965-66 to 4 per cent in 1974-75, we get net growth rates of crop production of 6.2, 8.8 and 7.5 per cent per year. Since population increased at the rate of 3 per cent per year between 1965 and 1975 per capita increase in crop production would range between 3.2 and 5.8 per cent per annum, an impressive rate of progress.

We have no data on the increase in milk production for the sample households; but as noted earlier, for the taluka as a whole, the milk yield per buffalo showed an increase of about 19 per cent over the period of 9 years or by about 1.95 per cent per year. These are official data. Buffalo population increased at the rate of 5.6 per cent during this period. Since buffaloes are the main source of milk, the quantity of milk production is estimated to have increased at the rate of 7.55 per cent per year. Overall increase in combined crop and dairy production per capita would range between 3 and 5 per cent. Relative to the performance of agriculture in LDCs in Asia and in India during the same period, agriculture in Matar had a better performance.

Pattern of Growth

Rate of growth, during the period under study was uneven over three regions of Matar. We have already observed that Navagam region had initially a high level of irrigation but further addition was small. In the other two regions, the initially low level of irrigation was followed by a more rapid expansion during the decade. The introduction of HYV of Paddy and Wheat made a perceptible impact on patterns of growth. *Bajra,* a staple food of the region, could barely maintain its relative importance. Cotton, an important cash crop, lost much of its importance. A combination of new technology and expansion in irrigation favoured growth in relatively less developed regions of the taluka. Secondly, and more importantly, the Green Revolution as such seems to have contributed marginally to the growth in Navagam. As already observed Navagam region had reached high levels of paddy yields prior to 1965-66, paddy variety *(Tinpankhali)* grown in Navagam before 1965 was known for its quality and commanded a price premium in markets even beyond the state boundary. Navagam peasants took to new varieties of paddy and continued to devote the same care and resources which they devoted to *Tinpankhali* in the past. New varieties brought them little gain by way of yields. Navagam still has the highest paddy yields. But inter-regional differences in respect of paddy productivity have however narrowed.

Irrigation

It can be asserted, on the basis of Matar experience that canal irrigation, though necessary, is not sufficient for the growth of crop production. When aided by conjunctive use of surface and subterranian water, a shift in cropping pattern and increased cropping intensity can be achieved. Winter crop of wheat following paddy requires dependable water which can be supplied by canal together with wells and tube-wells and worked with pumps and electric motors. In Matar taluka the number of pump-sets worked by oil engines and electric motors increased from 339 to 1068 between 1965-66 and 1974-75, as shown by the taluka records. Besides, they helped to increase the supply of assured water. Besides, the new paddy seed, which reduced crop duration provided an opportunity for a second crop which could be exploited with the help of wells and tube-wells, electric motors and oil engines. While Navagam experienced a change-over from paddy-fallow to paddy-wheat crop pattern, other two regions could skip the first paddy-fallow stage and enter intensive cropping stage by replacing *bajra*-cotton-crop-mix by paddy-wheat sequence.

It would have been interesting to measure the slope of the adoption curve for different sections of cultivators, and for all newly introduced practices. The 1974-75 survey results show that the adoption of all new important inputs was found in different size groups of holdings. Use of fertilisers and HYVs, the degree of cropping intensity or even the cropping pattern did not noticeably differ among big, medium or small farmers (Chapter 5, Table 5.1 and Chapter 7, Tables 7.2 and 7.3).

Energy Substitution and Polarisation

Use of bullocks for irrigation diminished over time with the spread of oil engines and electric motors. Canal irrigation required little energy for water-lifting. Substitution of bullock energy by other mechanical or electrical (or oil fired) water-lifting devices was completed in Matar before 1965. Between 1965-66 and 1974-75, the spread of canal irrigation and of conjunctive use of canal and lift irrigation made it necessary to complete land preparation for the second crop in a relatively very short period. This new task was performed by tractors. Tractors replaced bullocks for some of the operations related to crop production and marketing. In 1975, bullocks contributed 48 per cent of the total motive power used on small farms (0 to 5 acres) and only 10 per cent on big farms (25+ acres). For all farmers taken together, tractors accounted for over 70 per cent of the total energy used for traction. Substitution of energy sources from animal to mechanical and to oil or electrical, gave rise to market for services of indivisible productive assets. The separation of ownership from use of these assets led to a new type of polarisation of the structure of agricultural economy. This polarisation has been noticed in Matar which would be typical of all regions undergoing transition for modernisation. In the new situation the small and marginal farmers however become increasingly buyers of the services of energy sources in the same manner as tenants 'buy' the services of land in the lease market and the big farmers become rentiers.

Before 1965 the canal irrigation was predominant only in the Navagam region. In other regions well irrigation was dominant. As wells fitted with pumps energised by oil or electricity could cover a wide area, there developed a market for water supplied by lift irrigation.

Between 1965 and 1975, yet another market has developed. With expansion of irrigation and increased cropping intensity, tractors helped undertaking timely preparation for second crop. The number of tractors increased from 15 to 22 during 1965-66 for the families of cultivating households in the sample. The cultivated area per tractor worked out to below 200 acres, in intensity much above the national average and comparable to that of an advanced region. Out of 22 tractors only one belonged to a small farmer (0-5 acre size group). While ownership of tractors was thus concentrated among relatively bigger farmers, the extent of tractor use differed less among farmers in different size groups. The small farmers obtained from tractors 52 per cent of total motive power (mechanical and animal together) used. For medium and big farmers, tractors accounted for 75 to 90 per cent of total motive power.

The differences among farmers of different size groups pertain to hiring in of motive power. For small farmers hiring-in accounted for 58 per cent of total motive power but for big farmers only 15 per cent. Tractors had a large share in hired-in motive power. Typically small farmers hired 82 per cent of the tractor hours used by them. Big farmers also hired tractor services, but the hired-tractor hours accounted for only 17 per cent of the total tractor hours used by them. Dependence either regarding lift irrigation water or tractors on hiring is not uncommon among different size classes of cultivators. Thus dissociation of use from ownership of production resources develops; it is facilitated and encouraged by irrigation. Though as a consequence there could be increased landlessness and increased dependence on market for the services of the productive assets not owned by producers there is no adverse impact on income distribution and general well-being. This is the observation suggested by Matar resurvey.

Marketisation

Two developments – dissociation of ownership from use of resources and marketisation – are contrasting developments. Marketisation denotes increased participation in market transactions at both ends – inputs and output. Total paid-out costs for farming activity accounted for 34 per cent of the value of crop output in 1974-75. They do not include rent of land and maintenance costs of housing – two major assets. Charges for canal irrigation do not cover the cost of operation and maintenance, interest and depreciation. These costs are borne by the State. Full charge of canal irrigation would raise the share of paid-out cost. Two major items in paid-out costs are hired labour and chemical fertilisers, their shares being 28 and 20 per cent respectively. Bullocks accounted for only 2.2 per cent of the paid-out costs because of the nearly total disappearance of the market for bullock services (Chapter 7, Table 7.8). Of the total value of crop output, mainly food-grains, value of sales amounted to over 60 per cent in 1974-75. Though co-operatives have made demonstrable progress in respect of credit and dairy their involvement in crop sale in low. Data show that traders dominated, not even paddy millers, the former had a share of 75 per cent in the crops sold in the market in 1974-75. In Matar, the percentage of value of crop sales to crop out-put increased with the size of holding. However even the small farmers had a sizeable marketed surplus. Crops sales accounted for 53 per cent of total crop production in their case. For big farmers, the percentage of market sale was 73.

Regarding inputs, we find a different situation. Percentage of purchased inputs to total inputs was invariant with the size of holding, it ranged between 20 and 23 (Chapter 7, Table 7.8). Market has penetrated fairly rapidly even the conventional family production activity namely, dairy production, the purchased inputs accounted for more than half of the value of milk sales (53.3%) and milk sale accounted for 70 per cent of the fluid milk output. The dairy co-operatives have made a sizeable inroad, they had a share of 65 per cent of the sales of milk and 69 per cent of the sales of all dairy products. Paid-out costs include green and dry fodder and concentrates.

The dairy expansion in Matar is indeed rapid. But no link can be established between the extension of irrigation and the expansion of dairy production. Both are rather independent, exogenous economic phenomena; irrigation expansion is the result of a state decision for investment and the dairy co-operatives are the product of State encouragement to a district level initiative. A tenuous link can be traced in the form of irrigation facilitating the growing of green fodder. Green fodder growing though important is still in initial stages, it occupies only a small portion, estimated to be 0.1 per cent of total cropped area.

Marketisation has a direct and deeper impact. When a producer enters a market to sell his produce mainly crops (or even milk), his exposure to the market is limited. For crop production is exposed to market forces for crop sales once or twice a year. And as the traditional farmer was constrained by inputs available mainly from within the enterprise, or at best from within a limited geographic area (say a village), he tried to 'maximise production with the available resources'. Participation in market may still involve monetisation, i.e., selling produce for cash and buying with cash items of consumption. In an extreme case a producer may sell his entire produce to the market and buy his entire requirement of consumer goods from the market and the latter may include foodgrains. In this sense many labour households and small farmers already had 'monetised economy'. For instance many labourers earned wages in cash and bought grain from the market and small farmers sold their produce and obtained their requirements through purchases. Not infrequently sales could involve distress and purchase many involve peril.

In a traditional economy, dairy activity was secondary, primary being maintenance of bullocks. Of the available fodder supply bullocks are a major claimant. Milk production and milk sale fluctuated seasonally and surplus supplied to market varied more widely. Selling surplus to the market when available, limits market involvement. In Matar, dairy production is undertaken now mainly for market, with the help of sizeable quantity of purchased inputs.

Marketisation implies frequent exposure to market. When farmers start using inputs purchased from the market, exposure of the producer to the market force becomes fairly frequent. Further, every time the producer enters the market to buy an input brought to the market from the economic system that spanned beyond his immediate geographic surroundings, he would consider whether the transaction is worthwhile, and whether the expense incurred will bring adequate rewards. Such an exercise, done frequently inevitably changes his objective function from that of 'product maximisation' to 'profit maximisation'. The changed objective makes him less resistant to a change, he accepts a technological change embodied in modern inputs – improved seeds, fertilisers, pesticides – with much less resistance. He weighs benefits against costs and if the former are ahead, he undertakes a willing trial and a successful trial induces adoption of more innovations in rapid succession. It is in this sense we can state that marketisation influences economic behaviour and erases all major behavioural differences among farmers.

There is one more point. If a new technology involves 10 per cent probability of loss and 50 per cent probability of doubling income (and 40 per cent probability of 'no loss' 'no gain') for all size groups of farmers, the big and small farmers might react to the adoption of the given change in the same manner. It is this process which is implied in the observed phenomenon in Matar.

Increased involvement in market does not necessarily imply economic deterioration, though it may involve status retrogression, i.e., loss of control over resources, especially land. The distinction between status retrogression and economic well-being takes us to the discussion of equity.

Equity : Some Evidence

Increased demographic pressure could split some households. We call the households that branch off from the main ones as 'satellites'. The satellites as well as the original households have their land holdings diminished in size each time household splits and divides its land holding. Each time subdivision takes place it brings about a downward movement of the household on the economic status ladder. The satellites of those in the lowest size class would in all probability join the class of landless wage-earners. If there were no other forces working one would expect demographic forces to reduce the average size of holding and improve equity of land distribution among land holders. Consequently we expect the gini coefficient to decline and the proportion of wage-earning households to increase.

Land is an important asset. Its equitable distribution among cultivating households is desirable. We found that population expansion was not accompanied by reduced average size of holding between 1929-30 and 1965-66, also the increase in population between 1965-66 and 1974-75 was accompanied by only a marginal decline in the average size of occupied holding. The gini ratio of land operated by farm households declined from around 0.44 in 1965-66 to 0.41 in 1974-75, while it had increased from 0.38 to 0.45 over the preceding three decades. We have not distinguished between dry and wet areas; both are counted as equal in terms of acreage. The gini ratio may show a further decline if irrigation is given some weightage, since lower size groups have larger percentage of area under irrigation. The changes in gini ratio suggest an important question. Between the two namely, (1) resource base which in present case includes irrigation in a big way and the level of production technology, and (2) socio-economic forces, particularly demographic, which one would have a greater impact on the structure of land holdings? We should add a third factor namely, land reforms which had nearly frozen lease market and considerably reduced market for sale and purchase of land as an asset. The period of a decade may not be adequate to answer all these questions with certainty. A trend may however be indicated. It seems that while between 1929-30 and 1965-66 the technological base together with resource base moulded the structure of land holding, in the recent decade, the operation of the technological factor was accompanied by land policy which attempted to halt land concentration. During both the periods – the demographic forces had led to increased landlessness in agriculture (agricultural proleterianisation). While during the three decades prior to 1965-66, the process of increased landlessness was unchecked, during the recent decade it was a result of dynamic forces generating movement along occupation status scale both upward and downward. If opportunities were generated for employment outside agriculture on a sufficient scale, the process of increased landlessness within agriculture would have been defused to a large extent. To obtain greater insight into this process namely the dynamics of growth, we analyse distribution of assets (Chapter 11, Table 11.1).

Assets

In the context of growth and equity we reproduce below the gini ratios given earlier (Table 11.5). Between 1965-66 and 1974-75 we observe a change in equity in output and assets. Expansion of irrigation and associated technological change have helped reduce inequity both in output and in assets.

	1965-66	1974-75
Operated land (acres)	0.4600	0.4300
Operated land (value)	0.3837	0.3480
All Assets (value)	0.4317	0.3246
Assets (excluding land in terms of value)	0.4322	0.3774
Assets (excluding land + houses in terms of value)	0.4084	0.3274
Gross value of agricultural produce	0.4618	0.2724

Far more important was the reduction in the gini ratio of output, a decline which suggests rapid progress to equality. We may take a cautions view about production, since both 1965-66 and 1974-75 were abnormal years weather-wise. Changes in distribution of assets would suggest more firmly the underlying trend. Several gini ratios, of the value of land (alone), all assets, or assets excluding land or excluding land and houses, show that inequality has declined, and the decline was more than marginal. Further the decline in the inequality of assets distribution during 1965-66 – 1974-75 has changed the inter-asset relation. In 1965-66 gini ratios of land (area), all assets, assets excluding land, and crop output were within a small range. In 1975, their range increased because of sharp decline in gini ratio except that of land. In 1974-75, inequality in land (area) was the highest, that in all assets was the lowest and that for assets excluding land and houses was only marginally above that for all assets. Thus, by implication, during 9 years intervening between 1965-66 and 1974-75, accumulation of durable production assets by small asset holders was at a faster rate than that by others, despite the fact that mechanisation was apace during the period. The small asset holders expanded rapidly their animal herd size and added assets required for well irrigation (like diesel engines and electric motors).

Between 1965-66 and 1974-75, the relative magnitudes of inequality of land, other assets and output underwent noteworthy change. In 1965-66 output inequality was greater than that for distribution of land (area/value) as well as that in total assets whereas in 1974-75 the opposite was the case, the inequality of production was the lowest. Perhaps the public investment in irrigation has contributed to this change and made it possible for small producers to expand production relatively more rapidly and withstand bad weather better.

Over the nine years (1965-66 to 1974-75), the value of assets increased in real terms at an annual compound rate of 4.84 per cent, the investment (excluding the public investment in irrigation) formed around 25 per cent of net income (or 12 per cent of gross value of output). The investment was financed largely out of own resources of farm households. As a result debt was less burdensome. The burden of debt declined, relatively more for small farmers.

Consumption

A broader perspective regarding growth and equity can be obtained by considering all occupation groups together. For such an exercise we have considered changes in the level of living as well as expenditure on consumption.

Between 1929-30 and 1965-66 there was a marginal increase in per capita consumption expenditure in real terms for all households. For agricultural labourers and non-farm households, except traders, per capita consumption expenditure in real terms had shown a decline. Among cultivators, small farmers had a small drop in expenditure in real terms. We had limited data regarding quality of living as indicated by the quality of food and the durable assets (Shah, 1974, pp. 178-216). These data suggested that the strata that had experienced a decline in per capita expenditure between 1929-30 and 1965-66 seemed to be better-off when durable assets were taken into account. This was true of the non-farm families and farm labourers also.

For assessing the change during the period between two resurveys, we have consumption expenditure data for 1965-66 and 1974-75 (in current prices). We have to adjust these data for price changes using an appropriate deflator. The available cost of living index relates to agricultural labourers (COLI) in Gujarat as a whole. For comparison of consumption expenditure in 1929-30 with that in 1965-66 we had used paddy prices as the deflator. Between the two the former showed an increase of 135 per cent and paddy prices showed an increase of 67 per cent.

The per capita consumption of the sample households showed a decline in real terms by 15 per cent if COLI is used as the deflator. If we use the paddy price index as a deflator, we would get an increase of 19 per cent in the per capita consumption expenditure in real terms over the decade. Despite the problem of deflator we get a few firm findings. Firstly, whatever the deflator used, paddy price or cost of living for farm labour for Gujarat, we find that consumption expenditure of agricultural labourers, small farmers, and all non-farm occupation groups (except salary and wage earners) showed an increase in real terms. The salary earners showed an increase in real per capita consumption expenditure, on the basis of paddy price deflator only. The cultivators with more than 10 acres showed a decline in real per capita consumption expenditure irrespective of the deflator used (Chapter 12, Table 12.1). The increase in relevant cases was substantial if the paddy price deflator is used, and marginal if the cost of living deflator is used.

If we rearrange households in decile groups, for 1965-66 and

1974-75, and compare the average per capita consumption expenditure, decile by decile, we find, firstly, there was an increase in real consumption expenditure for all deciles (on the basis of the paddy price deflator) and secondly, the extent of increase was larger for lower deciles and the rate of increase declined monotonically (Chapter 12, Table 12.2).

Budget Shares

Improvement in the standard of living can be judged also on the basis of changes in budget shares. The budget shares suggest some improvement in the quality of consumption of food insofar as the share of non-grain food items has increased for all occupation groups except the farm households with holdings above 25 acres and the group of miscellaneous occupations categorised as 'others'. The share of expenditure of foodgrains had declined for all occupation classes except 'others'.

Consumer Durables

On the whole we find that consumer durables reflect improvement for all occupation groups. The aggregate index, which is a simple unweighted average of percentage of households reporting possession of different items of consumer durables, increased for all occupation groups taken together, from 27 to 40 or by about 48 per cent. This indeed it a steep increase. Such aggregate index increased by 72 per cent for farm labourers, 177 per cent for small farmers, 97 per cent for 5-10 acre group, 48, 47 and 23 per cent for size groups above 10 acres. Among non-farm occupation groups, aggregate index of consumer durable assets increased by 49 per cent for artisans, 61 per cent for miscellaneous occupation group, 56 per cent for salary earners and 27 per cent for traders. The above account is a testimony to an alround improvement of well-being of all occupation groups (Chapter 13, Table 13.1). We also found that ownership of prized items like brass and stainless steel utensils increased even among low income families.

Improvement for different occupation groups is impressive judged on the basis of individual items. Take the instance of huricane lantern, percentage of households in possession of it, increased, between 1965-66 and 1974-75, from 54 to 83 for 0-5 acre group (small farmers), and from 83 to 86 for 15 to 25 acre group (large farmers), from 37 to 61 per cent for labourers, from 73 to 86 per cent for artisans, from 57 to 68 per cent for salary earners (they include village peons, primary teachers), from 48 to 67 per cent for miscellaneous occupation group, while big farmers (25 acres and above) and traders showed a decline from 85 to 71 per cent and 93 to 87 per cent respectively. The latter two groups had improved their position regarding petromax and electric light. In the case of big farmers, percentage of households possessing petromax increased between 1965-66 and 1974-75 from 32 to 43 and that of households having electric lighting from 15 to 36. For traders those having electric lighting increased from 37 to 50 per cent. For agricultural labourers, items that showed impressive increase in terms of percentage of households possessing them were stainless-steelwares, glasswares, chinawares, wooden beds, huricane lantern, primus stove, wrist watches, fountain pens (including ball point pens – a sign of increased literacy), umbrella and suit-cases. The above list could be expanded to include radio, bicycle, wooden furniture and steel furniture for small farmers, and it would be further expanded to include thermos flask, time piece, and wall clock for medium farmers, and further still to include, electric fans, rain coats, motor bike for big farmers. Nearly 50 per cent of big farmers possessed radio against 12 per cent of small farmers and 3 per cent of farm labourers. Similar pattern emerged for different non-farm occupation groups.

Somewhat greater details presented above for consumer durables show convincingly that, on the whole, over a period of time, keeping weather effect out as reflected in current consumption, the material well-being of all occupation groups seems to have improved. This is reassuring, as it shows that development generated by irrigation percolates perceptibly to lower income groups.

We may sum up our empirical observations with following inferences:

1. Growth in farm production induced by irrigation and improved technology, leads to modernisation and marketisation of rural economy.
2. Such development seems to have definite percolation effect, the lower income groups to derive benefit of growth and that the extent of such benefit in all probability is greater for them, as judged by reduced inequality of production, assets, consumption, and improved consumption of durable items.

An Explanation and a Hypothesis

We observe two phenomena that need explanation. We observe that the growth of farm production has caused improvement of standard of living and this improvement has percolated to lower economic strata. We also observe that the percolation is not large enough to banish poverty (misery) or low living standards completely, the process of growth has bypassed some sectors of households. These two phenomena of (i) percolation and (ii) bypassing existing side-by-side need an explanation.

Our explanation is in the nature of a hypothesis. The extent of coverage of cultivated area by access to irrigation is a crude measure since not all areas have equal dependability of irrigation system to meet water needs. The gap between the need and availability widens or narrows depending on the natural supply of water by rainfall. As an hypothesis it can be suggested that where irrigation is one of the major production resources the production fluctuations widen. While irrigation reduces dependency on rain, dependable supply of irrigation water encourages introduction of crops in production system, which are more weather sensitive especially to temperature, moisture etc. Some of the HYVs of wheat and paddy are known to be photo sensitive and have greater proneness to attack by pests and diseases.

Irrigation has induced modernisation of technology and through it greater market ties regarding inputs and output. In a system such as this, expectations, ability to bear risk, access to inputs, access to credit and investment abilities assume major importance. As a hypothesis, it can be suggested that these abilities are only loosely dependent on size of land holding, operators with the same size of holding may evince unequal abilities to face market. We have not analysed in detail the linkages among production, investment and risk bearing abilities. We found in Matar that access to irrigation was relatively greater for small farmers and small farmers showed relatively greater progress in production (or ability to withstand the weather inclemency).

Percolation process seems to have reached lower income strata including agricultural labourers. As a group they have experienced increase real consumption, larger possession of consumer durables. We suggest as a hypothesis that when irrigation induces a major change in cropping pattern, the shift would be in favour of crops with greater labour absorption, which is further increased with increased cropping intensity. Paddy is acknowledged to have larger labour absorption capacity than *bajri* (pearl millet). Paddy, followed by wheat, absorbs more labour than two-season cotton crop. Use of irrigation together with a change in cropping pattern and increased cropping intensity tends to –

1. encourage greater overall demand for labour.
2. narrow inter-peak time intervals.
3. raise both peak and off-season employment levels.

Consequently in Matar, the demand for labour increased steeply which in turn induced greater in-migration of labour, mainly seasonal. Migrant labour contributed 21 per cent of hired labour and constituted 14 per cent of total labour in 1974-75, against 2 per cent in 1965-66 as inferred from field data.

Despite major increase in labour in-migration wage-earning of labour in real terms tended to increase. The increase could be inferred to be substantial if wages for *bajri* operations in 1965-66 are compared with that for paddy operations in 1974-75, the increase in real terms would range between 50 to 70 per cent. For paddy operations as such the increase in real wage between 1965-66 and 1974-75 would be marginal.

Increased employment, together with increased real wages despite in-migration of labour made possible income increases for labour households. Our observation is that marketisation of agriculture between 1965-66 and 1974-75 changed also the character of labour market. Between 1965-66 and 1974-75 labour households reporting land holding including those with land holding of 5 acres and above declined. Besides, percentage of labour households reporting other occupations as subsidiary also declined. Thus in 1974-75 most of the farm labour households depended mainly or exclusively on agricultural wage employment. Besides, exchange labour and annual labour were found to be almost on a way out. The labour market thus consisted of wage-earners, employed on daily basis and yet they showed improvement in their levels of living. Our hypothesis is that rapid increase in irrigation increased employment opportunities that led to integration of labour market beyond the boundaries of the local region. Since immigration involves cost of information and migration (including costs of agents that bring outside labour) such costs raised the opportunity costs of bringing in migrant labor and thus permitted a rise in real wages of local labour. The net effect of mechanisation in

Matar taking big and small farmers together, was not labour substituting, it saved mainly animal energy power.

Integration of labour market induced inmigration on a large-scale and inmigration had what might be termed as 'floodgate effect' on labour market. When migrant labour comes in a big way, market would exercise option in favour of able and efficient labour. In contrast when peak demand is to be met only by mobilising local labour, all available labour irrespective of production capacity of participation individual would get employment.

Situation regarding agricultural labour is typical and it illustrates the process of percolation of benefits of growth to the lowest stratum. The specific growth promoting scenario of increased resource base through public investment in irrigation and consequent change in farm production technology seems to augment the labour demand and therefore permit the benefits to percolate to the lower strata in terms of increased earnings which would be the combined result of increase in the employment and their wage rate. Such benefit spills over beyond the boundaries of the region and labour from other areas also benefit.

Market integration not only allows percolation but also the spread of benefits to sectors other than agriculture. We have noticed increased current consumption and purchase of durables, besides increased use of purchased inputs. Indeed these forward-backward linkages build bridges for benefits to flow freely over them to other regions and other sectors. Such integration has a cost. Bridges are built at a cost, such costs determine how much and how far the benefits would percolate. The percolation of benefits to other classes in rural areas is illustrated by the fact that on an average there were in 1974-75 4 retail outlets per village (and number subsequently increased to over 10 in 1982-83). This number irrespective of the size of enterprise or income is suggestive of the growing trade sector in rural areas. The new retail outlets provide employment options to those engaged in agriculture and village crafts that are on the decline. Matar has also seen the rise of small industrial establishments of two types: (i) those linked with processing of expanding farm production and (ii) those not connected with farm production but reflecting the growing spirit of enterprise in rural areas. The former has resulted in an increased number of rice mills, while the latter has been evident in the form of the setting up of tape-manufacturing, machinery spare parts manufacturing etc. The risk capital may have been drawn from the growing surplus in agriculture. The marketised agriculture becomes capital demanding besides being labour demanding. Land being invariant in area, public irrigation promotes systematic use of land-augmenting capital. These include such current inputs as fertilisers and such durable assets as oil engines, electric motors, pumps, tractors and other machines. The rate of growth of physical durable production assets, 4.34 per cent per annum, implied a fairly high rate of ploughing back of income, about 25 per cent. Such high rate of ploughing back of income for productive investment in physical capital however was confined to only 36 per cent of all cultivators. While the increased demand for labour would benefit most of the farm labour households the increased demand for physical capital could benefit only a few, mostly enterprising, farmers.

An element of enterprise required for farmers to benefit from the growth process is demonstrated further by adoption of improved practices. Three major practices-irrigation, fertilisers and HYV seeds together were used by 44 per cent of farmers in 1974-75, compared to nearly 84 per cent who used irrigation, for more than eight years after the Green Revolution. The percentage of adopters was invariant with size of holding. The extent of adoption, i.e., coverage of area and intensity of use, i.e., dose applied, vary. Among adopters, those with small holdings used larger doses on wider area of their fields.

This element of enterprise required for growth, resulted in dissociation of size of operational land holding from growth process benefits.

Against nearly 96 per cent of the farm labourers having full employment at higher wages, one-third to two-fifths of cultivators participated in the process of growth. The producers benefiting from growth were evenly spread over size-classes of land holding – with middle size classes having some edge over others. The process like this produces dynamism and poverty and prosperity – beyond the confines of traditional occupational classes. With growth requiring enterprise and progressive outlook, the growth-generated prosperity percolates to lower classes including labourers.

In Matar, the capacity of producers in the two regions to respond to growth impulses in a short time has made possible what was feasible namely growth with equity.

Perspective

The experience of Matar cannot be moulded in anyone specific model of growth and distribution known to the profession. The process in this small region was initiated by public investment in irrigation on a large scale, followed by complementary investment by individual producers, aided by market development for dairy product, an exogenous factor, and facilitated further by embodied technological change represented by improved seed and fertilisers. Growth rate between 5 and 6 per cent, by nature labour demanding has demonstrated, even with increased proletarianisation, an improved income and asset (and production and consumption) distribution.

The experience of Matar has elements of most of the theoretical models. The growth was labour demanding, it had a technological improvement component, benefits distribution would corroborate efficiency proposition, but findings for 1965-66, 1974-75 would go counter to widely documented results, the growth from low levels of incomes has resulted in reduced inequality of income and assets.

What is suggested by Matar experience is that a change in technology has to be substantial, if it has to cross all institutional hurdles and spread benefits to the lowest class and beyond the regional boundaries. The experience is very happy but could be very special too. Crucial to the dynamism generated by high growth is the process of marketisation, a wider experience than commercialisation.

REFERENCES

Deolalikar, Anil (1985) "Direct and Indirect Effects of Irrigation, The Case of Matar Taluka in Gujarat State — India 1965-66 to 1974-75" (Mimeo).

Kumar, B.L. (1985) *Production and Transfer of Goods and Services in Matar Taluka*, The Gujarat Institute of Area Planning, Ahmedabad. pp. 28-36.

Patel, B.B. (1985) *Impact of Irrigation on Labour Absorption and Labour Market Adjustment in Agriculture in Matar Taluka, Ahmedabad*, Gujarat Institute of Area Planning.

Patel, N.M. (1984) *Effect of Rainfall and Rainy Days on Yields of Major Crops in Matar for 1965-66 and 1974-75*, Ahmedabad, The Gujarat Institute of Area Planning.

Shah, Vimal and C.H. Shah (1974) *Resurvey of Matar Taluka*, Bombay, Vora & Co. pp. 178 to 216.

Thorner, Daniel (1980) *Shaping of Modern India*, Allied, pp. 325 to 328.

Vaidyanathan, A. (1982) *Role of Bovines in Indian Agriculture: A study of Size, Composition and Productivity*, Trivandrum, Centre for Development Studies.

INDEX